MORE PRAISE FOR *MAKE, THINK, IMAGINE*

'John Browne makes a compelling argument about the power of technological progress and its ability to bring prosperity to the global community. He expertly backs this central argument with in-depth historical research, interviews and his own personal experience to provide a blueprint for future global progress underpinned by the spirit of innovation. It is also a long-overdue tribute to the importance of the engineer in our society' Lord Foster, Founder and Executive Chairman, Foster + Partners

'*Make, Think, Imagine* is for anyone curious about how our world was created and asks deep questions about our possible future. Lord Browne's beautifully written and easily accessible work breaks down the forces behind our innate desire to create, to point us in the right direction: an inclusive future that gets us a greener, healthier and safer planet' Tony Fadell, iPod inventor, iPhone co-inventor, Nest founder and Future Shape principal

'A fascinating tour de force from someone who has walked the walk but has gone well beyond his own experience to explore the centrality of engineering to our lives and to meeting the challenges of the future. As someone unable even to change a plug, I surprised myself in enjoying a book about engineering so much' Alastair Campbell, writer, campaigner, strategist

'*Make, Think, Imagine* is a very thoughtful, deep journey through our fast-moving and fast-changing evolution, describing how engineering has and will continue to change the landscape of our civilisation' Ratan Tata, chairman emeritus of Tata Sons

'Engineering technology is the necessary catalyst for progress in humanity. It's magic!' Professor James J. Spilker, Jr., Stanford University, co-inventor of GPS and winner of the Queen Elizabeth Prize for Engineering 2019

'A fascinating overview of the vital roles played by engineers in every aspect of modern life, and an important and optimistic vision of how engineering will be essential to provide a prosperous and healthy future for humans living on our densely populated and increasingly polluted planet' Sir Mark Walport, Chief Executive of UK Research and Innovation (UKRI) and former Chief Scientific Adviser to the UK Government

'A compelling tale of humanity embedded in an engineered world: past, present and future. John Browne opens our eyes in conversation with other experts with this story of improving lives' Dame Sally Davies, Chief Medical Officer for England

'In an era of technological advancements, many argue that we are moving too fast and feel threatened by innovation. This remarkable book takes an insightful dive into the past, drawing out some of history's pivotal engineering innovations and demonstrating how they helped civilisation solve problems more efficiently. Through his powerful analysis of our past successes and failures, John Browne delivers a clear argument for a brighter, safer, healthier future fully embracing innovation. This book is a must-read for everyone with an interest in improving human life and overall wellbeing' Lord Ara Darzi, Professor of Surgery at Imperial College London and pioneer of robot-assisted surgery

'At a time when gloom and pessimism abound, *Make, Think, Imagine* provides a welcome blast of good news. John Browne, in a compelling and insightful book, shows that we have the creativity, the tools and the know-how we need to build a brighter future. If we design and use it responsibly, engineered technology can and will make us healthier, safer and more free, wherever we live. We need to grasp the moment and have the courage to see it through' Sir Jeremy Farrar, Director of the Wellcome Trust

'The public is quite unaware of the impact of engineering, and John Browne's summary, I believe, is accurate – modern civilisation is primarily based on engineering' Professor Brad Parkinson, Stanford University, co-inventor of GPS and winner of the Queen Elizabeth Prize for Engineering 2019

'In a time of doubt and pessimism John Browne presents the rational, evidence-based case for optimism and belief in the possibility of human progress' Nick Butler, energy commentator, strategist and visiting Professor at King's College London

'In our time of huge environmental challenges John Browne's book is a joyful reassertion of the phenomenal problem-solving capacity of human beings. By arguing that engineering is the fundamental driver of all progress he successfully asserts that there is the real potential for society's problems to be solved and for the world to move forward again' Thomas Heatherwick, designer

'Lord Browne has brought together the insights of some of the finest engineering minds on the planet in this elegant love poem to engineering across the ages ... A rattling good read' Vivienne Parry OBE, writer and broadcaster

'This inspiring, ambitious book celebrates the contributions engineering has made to progress in the past, and explores thoughtfully the role it can play in shaping the future' Professor Diane Coyle OBE, University of Cambridge

'A call to intelligent and sensitive action. John Browne's directory of a sustainable future gives us hope and the outlines of a roadmap to get there' Sir Antony Gormley, sculptor

'Science and technology have transformed our world. This book argues persuasively that our civilisation is dependent on them, and if the world is to prosper then we need to harness both effectively for the future of humankind' Sir Paul Nurse FRS, Director of the Francis Crick Institute and winner of the Nobel Prize for Medicine

'Like the engineer he was trained to be, John Browne gathers wisdom from over a hundred innovators, weaves into it his own diverse experience and reveals a realistic and optimistic fabric of the future. We can get there and John tells us how' Vint Cerf, Internet pioneer

'Have you ever wondered what it would be like to wander the globe, looking behind the curtains of the world's greatest collections and achievements and discussing them with the people that created them? Browne takes us on a personal and technological tour de force of breathtaking expanse … What sparks from every page is the human passion, be it through art or engineering, to make – to fashion and have an impact on the world around us. It is a view that brings art and science back together; that shines a light both forward and back on the good and bad that technology can bring; but that is ultimately optimistic about what we can achieve. Browne rises above today's narrow disciplines to see afresh the world our hands have fashioned' David Halpern, author of *Inside the Nudge Unit* and CEO of the Behavioural Insights Team

'Browne has written a timely book that connects his deep knowledge of business, culture, history and science. *Make, Think, Imagine* is a much-needed antidote to the drift and pessimism gripping advanced industrialised democracies' Lionel Barber, Editor of the *Financial Times*

'Lord Browne has written an ode to the ways in which engineering has improved human civilisation, from how we communicate, build, use energy and move to how we explore the universe. *Make, Think, Imagine* is a joy to read, with interesting historical insights and a vision of a better future' John Hennessy, Chair of Alphabet (parent company of Google) and former president of Stanford University

'John Browne always makes me think. He is ahead of the curve' David Miliband, President and CEO of the International Rescue Committee and former UK Government Foreign Secretary

MAKE, THINK, IMAGINE

Engineering the Future of Civilisation

JOHN BROWNE

BLOOMSBURY PUBLISHING

LONDON • OXFORD • NEW YORK • NEW DELHI • SYDNEY

BLOOMSBURY PUBLISHING
Bloomsbury Publishing Plc
50 Bedford Square, London, WC1B 3DP, UK

BLOOMSBURY, BLOOMSBURY PUBLISHING and the Diana logo are trademarks of
Bloomsbury Publishing Plc

First published in Great Britain 2019

A catalogue record for this book is available from the British Library

ISBN: HB: 978-1-5266-0571-9; TPB: 978-1-5266-0570-2; EBOOK: 978-1-5266-0569-6

2 4 6 8 10 9 7 5 3 1

Typeset by Newgen KnowledgeWorks Pvt. Ltd., Chennai, India
Printed and bound in Great Britain by CPI Group (UK) Ltd, Croydon CR0 4YY

To find out more about our authors and books visit www.bloomsbury.com
and sign up for our newsletters

To my father, who told me to get a real job

The author's profits from this book will go to The John Browne
Charitable Trust, where they will be used for the education of
engineers and support of the arts

CONTENTS

1

Progress

We all have a deep-seated urge to improve our lives. We want to look after our health and well-being. We want to make the lives of our families better. We want to enhance the future of our nations. We want to travel and to communicate. We want to know that our voice is heard. And, driven by an innate human instinct, we want to make things and shape the world around us. The most effective way to do all these things is through engineering; without it we cannot make progress.

There is an engineer in every one of us, but we fortunately do not require the skills or expertise of a professional engineer to tap into this aspect of our nature – contemporary technologies, such as the Internet and smartphones, give all of us access to an engineering mindset. Now, more than ever, we can use these technologies to solve the world's problems and shape the society in which we live. We are the stewards of that society, with views about how we want people to behave, how technologies and businesses should be regulated, how leaders should act and, crucially, how all this impacts our daily lives. We express our views through the values we espouse, the things we buy and build, and by those we elect.

Engineering is wrapped around all of us, like a protecting and life-sustaining blanket. Theory and academic debates have their place, but engineers are best known for their practical impact; while others talk and pontificate, they are out in the world, influencing and shaping it. If you look around, you will see a world made richer, freer and less violent by engineering. It may not be immediately obvious, but advances, from the first rough-hewn stone tools to the exquisitely poised qubits of a quantum computer, have stimulated every single important step forward.

Since I was born in 1948, at least twenty new vaccines have been engineered and produced, eradicating or limiting the spread of many crippling diseases.[1] The proportion of the world's children that die before they reach the age of five has dropped from more than one in five to fewer than one in twenty-five. In the world's richest countries, the infant survival rate is ten times better still.[2] During my lifetime, average life expectancy has increased by more than two decades.[3] The prime instigators of these advances have been the systems we have engineered to provide medicine, food, water, sewerage, energy and, in its fullest and most liberating sense, prosperity.[4] And whereas nearly three quarters of all people lived in extreme poverty in 1950, less than 10 per cent do today.[5] People are, on average, not just better off; they are also better informed and educated – global literacy rates have climbed from just 35 per cent to more than 85 per cent during the same period.[6]

Progress is precious and must never be taken for granted. It is under fire from those who have been ignored, those who crave the past, those who feel threatened by unconstrained globalisation and those who believe that things are changing too quickly. But stopping progress is impossible, and those who try to do so will certainly fail.

Progress happens when we connect ideas from different fields of human endeavour, mix them up, try things out, learn from our mistakes and try again; that is what is meant by 'trial and error', and it is the fundamental process that gives us progress. The Wright brothers invented the aeroplane by trial and error, and subsequent trials led to bigger and better aeroplanes. With the advent of the turbine, the jet aeroplane was born. The supersonic passenger aeroplane, Concorde, was next, before prototypes of the 'hypersonic' aeroplanes that might, in decades to come, take us between continents at unprecedented speeds. And while the biggest and fastest innovations may capture our imaginations, the products of engineering must, in the end, also be reliable and trustworthy – after all, no one wants to fly in an aeroplane that has only a 95 per cent chance of staying in the air.

But progress is about more than just functional rationality – it is a human activity and so it is also about beauty, art and the irrational. For example, the bullet train exists because Japan wanted to create a modern impression while hosting the 1964 Olympic Games. At the start of the twentieth century, gasoline-fuelled automobiles became the popular choice because they conveyed a more masculine image than the electric-powered alternatives. In London there is a bridge that, rather than lifting to let boats pass, rolls up into a ball. There is beauty to be seen in every great piece of engineering, whether a computer algorithm coded with a minimal set of instructions, the great array of telescopes in the Atacama Desert or the concealed flying buttresses of St Paul's Cathedral in London.

Even advances that we might not think of as technological depend heavily on engineering. The steady march of democracy would have faltered at the very beginning without the invention of reliable techniques with which to record laws and tally votes. How much difference could Martin Luther King Jr., Emmeline Pankhurst and the other great champions of humanistic values have made without the microphones, radio broadcasts and newspapers that amplified their messages and showed the world that there was another way? Without the engineered means to apply, distribute and pass it on to future generations, knowledge is impotent.

It is an exaggeration to claim that engineering was solely responsible for all these advances, but its critical role is too often overlooked. When it is done well, engineering produces innovations that allow us to solve our problems and improve our well-being, while expending less effort and cost. However, if we want to advance civilisation, we must strike a balance between the drive to innovate and the need to preserve a stable society. This was understood 500 years ago by the engineer and philosopher Georgius Agricola, who saw clearly that engineering could bring enormous benefits to all of society, but only if businesses employed its products in consultation with, and without the exploitation of, local communities. He summed this up by writing that 'good men employ them for good, and to them they are useful. The wicked use them badly, and to them they are harmful.'

That remains true today; the way people choose to use an innovation will determine its impact on society. But every engineered product will also generate its own set of consequences, both intentional and unintentional, as well as constructive and destructive. The same engineering that produces drones that deliver medicines to remote and disease-stricken communities also produces drones that are used in assassinations. Genetic engineering will cure some diseases, but it could also produce new pathogens. Opioid painkillers can relieve suffering, but can also cause addiction. Open communication and connectivity have allowed us to be expansive in our access to data and the use of it, but have also permanently weakened our ability to keep things private. Since the discovery and large-scale manufacture of penicillin, antibiotics have saved billions of lives, but their indiscriminate use has led to the appearance of drug-resistant bacteria, which if not eliminated, will cause great suffering. Hydrocarbon fuels have been the foundation of the great advances since the eighteenth century but using them produces greenhouse gases, which are dangerously altering the Earth's climate.

Progress is not delivered with an instruction manual spelling out the safe and responsible use of new inventions. Engineering is instead like a game of cat and mouse, in which innovators must continuously act to ensure that the intended consequences of their efforts outweigh the unintended ones. Engineered solutions will never be perfect first time, because mistakes and misuse are inevitable and every step forward has risks. Autonomous vehicles will create a revolution in convenience but, unless properly designed and tested, could kill more people than human drivers currently do. Medical advances will prolong life, but that will be for nothing if we cannot handle the dementia and loneliness that will become increasingly common with compassion and empathy. Robots and artificial intelligence will make life easier but could cause large-scale unemployment. Despite occasional and highly publicised failures, we must not succumb to the belief that we should slow the pace of innovation – if that happens, everyone will lose out. We must, however, think long and hard about how to react when things go wrong. The precautionary principle is not the answer, since striving to avoid all possible risks can halt or even reverse progress. For civilisation to progress, and for everyone to have access to freedom, learning and opportunity, we need innovators who can challenge accepted wisdom and make new things.

We therefore need a more sophisticated way of looking at the risks created by engineering advances, and that requires a shared belief in rational analysis and some consensus about what level of risk is acceptable. For example, Amnon Shashua, the founder of Mobileye, a company that develops the software for autonomous vehicles, believes that society will only accept automated systems that are a thousand times safer than human drivers. 'Dog biting man is something that we're all used to,' he explains, 'but man biting dog – or a machine killing a person – is something very exceptional.' It is also not right that one person's fear should be imposed on others; for example, enormous damage was caused by the irresponsible and unfounded claims that the measles, mumps and rubella (MMR) vaccine was harmful. It is essential, however, that we test every anti-intellectual attack on progress with compassion and care, ensuring that the likely risks and benefits of any advance are understood, before clearly communicating them. Innovators must strive to understand the hopes, needs and fears of society, but they can only do this if they marry their visionary insight with the expertise of others who have a deep understanding of the human condition. That is the best way to reduce the likelihood of myopic thinking and bad decision-making. On a practical level, it means that innovators must engage with a diverse group of people who have witnessed and learned from past failures – this is a powerful practice that must be employed. And those who are trusted to communicate the risks and benefits of an innovation are not necessarily governments or 'experts', but are just as likely to be role models, friends or family. Effective communication requires engagement with all agendas, and not simply with that of the communicator.

In the face of all this, it is unsurprising that some people believe that, when it comes to technological innovation, the bad outweighs the good. They view our historical failures to anticipate the consequences of innovation as evidence that what may at first appear to be in the public good will often turn out to cause devastation. That view is understandable, but it risks condemning us to the belief that humankind's condition is one of inevitable failure – a sentiment that has echoed across cultures and across the centuries. The media reinforce this negative belief by exploiting the fact that bad news sells, which plays to people's innate tendency to draw wider conclusions about the likelihood of bad events from their own immediate experience. This, in turn, fuses with our general tendency to expect losses more than gains.[7] In an era when people feel disenfranchised by unconstrained globalisation, are worried by unexplained technological progress and think change is happening too quickly, populist politicians have been able to stoke public distaste for progress by referring to a mythical past in which the world seemed better and more prosperous. The end result of all this is that large swathes of the public, across many nations, are firm in their belief that the world is getting worse.[8]

My mother would never have tolerated that sort of gloomy outlook. Having miraculously survived the horrors of Auschwitz, she felt that nothing good came from dwelling on the past and thought the best was always yet to come. Her strength and outlook inspired me so much that, when it came to earning a living, I wanted to solve problems that others had not yet even conceived of, and to help find practical solutions to humanity's most pressing problems – and this was why I decided to become an engineer.

Engineering naturally led me into business, since I realised that no solution was complete unless it resulted in something practical that humanity wanted. Thomas Edison apparently said that 'anything that won't sell, I don't want to invent. Its sale is proof of utility, and utility is success.' Engineering is like a head with two sets of eyes: one looks to the fruits of discovery, while the other looks to the demands of commerce and customers. The brain makes solutions, but it is only effective if it integrates all that it sees. As I experienced more of the world, I understood that there was far more to business than simply putting engineered solutions into the market – I came to see that, although engineering is a necessary and a vital start, it is not on its own enough to pull humanity out of barbarism and into civilisation. When engineering moves forward without giving sufficient thought to its long-term impact on society, it can halt or even reverse progress.

In the spring of 2018, the world realised that Facebook, a wildly popular and highly engineered product that allows people to share their experiences, may have become something of a social menace. As a result of its incredible success, it had become a repository of revealing information about the lives of over 2 billion people. Facebook is reported to have pursued relentless growth, of both its level of influence and its corporate profits, at any cost, a strategy that was revealed starkly by an internal memo in 2016, in which a senior Facebook executive wrote, 'Maybe someone dies in a terrorist attack coordinated on our tools. And still we connect people. The ugly truth is that we believe in connecting people so deeply that anything that allows us to connect more people more often is *de facto* good.'[9] Like Facebook, many companies, government departments and other organisations have realised the power and value of 'big data'. This has created a new industry, which grew by pursuing what is possible, without pausing to consider the potential damage its actions could inflict on people's privacy and trust. Leaders in this area have too often lost sight of the consequences of their actions, following an ambition summed up by the misguided Silicon Valley mantra, 'move fast and break things'. When little or no thought is given to the impact that engineering has on society it can cause great harm.

When society notices that something is wrong, it responds vigorously. For example, many citizens in the US and beyond who have become aware of the negative aspects of surveillance are reacting against Facebook, and the company will need to show that it is listening and that it can adapt its business model for the betterment of society. The dominance of Facebook, Amazon and Google and the like has some parallels with J. D. Rockefeller's Standard Oil Trust;

as valuable commodities that can be extracted from the world and used to create influence and wealth, oil and 'big data' have much in common. Eventually, it was a journalist, Ida Tarbell, who broke up the Standard Oil Trust. Her father had been forced out of business by Rockefeller, so she had a personal insight into the ways that aggressive and monopolistic business practices can hurt individuals and families. Tarbell single-handedly exposed Standard Oil's bullying tactics, changed public opinion and ultimately forced the US government to restrain Rockefeller's ambitions. Describing the oil magnate, Tarbell wrote that he 'has systematically played with loaded dice, and it is doubtful if there has ever been a time since 1872 when he has run a race with a competitor and started fair'.[10] Similarly, some people today are starting to question the fairness of the dice that the leaders of the largest technology firms play with.

In this book, I have gone both back into the past and forward into the future, to demonstrate that engineering is at the heart of all human progress. I have spoken with over one hundred of the world's greatest innovators, from surgeons to architects to computer and medical scientists. All of them display a blend of hopefulness and pragmatism, which seems to me the only viable alternative to the blind faith of the optimist and the fatalism of the pessimist. They speak eloquently about the huge benefits that engineering has brought to society, from cheap ballpoint pens to cloud computing, and from sewerage to satellite-based navigation. Engineering has shaped every aspect of our world, and it continues to drive progress across all fields of human endeavour. At its core, it is about the shared human urge to make things. I am sure that is why, at least in the West, people worry when they see domestic manufacturing facilities closing down and their goods being made elsewhere – such changes seem to diminish their country's standing in the world.

We urgently need to rekindle confidence in our ability to make progress, which is what I hope to do in this book. I will explain how the human urge to make things has generated great innovation that has, throughout our history, changed every aspect of the way we live. I will show how we can ensure that innovation has a positive impact on society, and how its unintended consequences and intended abuses can almost always be counteracted or prevented. I remain an optimist, because nothing can be achieved if we decide at the outset that we have failed. We need a clear-eyed belief in our power to shape the world for the benefit of all humanity. And that is exactly what engineering will do for us.

2

Make

HANDMADE

As I hold a stone hand axe in the palm of my right hand, the heavy object takes me to the heart of what it means to be human and demonstrates our basic belief that the world around us is malleable – we can use our own hands to cut, scrape and pound the world into the shape that we want it to be. In the 1990s I was a Trustee of the British Museum. One of the privileges of the position was being able to inspect what was in the stores, and it was there that I examined this hand axe. It was 40,000 years old, so relatively young as these things go, crafted out of semi-translucent, coffee-coloured flint in the elegant shape of a teardrop, and surprisingly warm to the touch. Whoever made it had a good understanding of its function; it fitted snugly into my palm, a keen edge tracing the perimeter. Modern-day butchers who have tried to use these ancient tools are surprised to find them just as efficient as well-honed steel knives. The hand axe was humankind's original handmade invention and, if longevity is any measure of success, it was also our best. The oldest stone hand tools, found in Kenya, are more than 3 million years old.[1] For countless millennia, all the work we did was fuelled by muscle power alone, and the hand axe was the epitome of technological sophistication.[2]

Not long after I held the axe, a friend took me to an auction of old watches in Switzerland. He told me he was keen to buy one made by Breguet; I had no idea what he was talking about, but I went with him to the pre-sale exhibition, and what I saw opened my eyes to a branch

FIGURE 2.1 To axe, or not to axe – shaping things 300,000 years ago, with a flint hand axe.

of bespoke manufacturing that I had never seen before. Here were masterpieces of ingenuity – small worlds of gears, springs, wheels and ratchets, all in the service of telling the time. Some were so elaborate that they were called 'grande complications'. I fell in love with the idea that such effort had gone into achieving a simple objective of measurement, which can now be achieved by an inexpensive digital watch.

These two objects, the hand axe and the watch, are both finely crafted small objects, but they are separated by tens of thousands of years of history. Despite their very different constructions, they have a surprising amount in common – they both demonstrate humankind's desire to use its manual dexterity and resourcefulness to make new tools. Across history, it is tools like these that people have used to make their lives better. The late Calestous Juma, Professor of the Practice of International Development at Harvard University, believed that making things is at the root of all of progress. 'Society advances because we make things. It's just so obvious,' he said with a chuckle.

FIGURE 2.2 More than just a time teller – a masterpiece of bespoke making that I watch on my wrist: my Breguet.

The watchmaker Abraham-Louis Breguet was a master of the handmade and introduced a series of innovations that revolutionised the reliability and accuracy of the pocket watch during the late eighteenth century. Napoleon bought three of Breguet's watches and used them to coordinate his army's actions on the battlefield. Throughout his long career, Breguet's quest was to perfect the pocket watch and, as a result, every one of his timepieces was unique. His most famous watch, commissioned by French queen Marie Antoinette in 1783, was so complex that it took forty-four years to complete and left Breguet's workshop only after both queen and watchmaker were dead. When it was finally finished, the Breguet No. 160 was a masterpiece of mechanical complexity. It kept track of the date (including the ability to accommodate leap years), chimed the hour, had a power reserve indicator, a stopwatch and even incorporated a thermometer.

MASS PRODUCTION AND THE MANUFACTURING MIRACLE
Breguet's approach to making was the antithesis of his contemporary, the French engineer Honoré Blanc. In November 1790, Blanc put on a dramatic demonstration for a group of politicians and generals. By

selecting components, apparently at random, from bins arranged in front of him, he quickly assembled several working muskets. He boasted that no one before him had 'seriously concerned himself with perfecting the firearm', but here 'every [component], without exception, has been thought about and discussed'.[3] France was in the throes of bloody revolution and on the brink of a series of foreign wars; it had an urgent need for affordable and reliable weapons. The crux of Blanc's sales pitch was clear; it did not take a highly skilled craftsman to build a musket – if the gun parts were designed by an engineer and machined to meet pre-specified criteria, they could be assembled by almost anyone. Lieutenant General Gribeauval, a key supporter of Blanc, wrote that the new approach would lead to a 'real and considerable reduction in the price of arms ... due to the infinite abridgement of labour costs'. And Blanc's approach delivered guns that were not only cheaper than the competition, but also far more reliable. By the time he died in 1801, his factory at Roanne in central France was turning out more than ten thousand identical muskets a year for Napoleon's armies. This was a new way to manufacture that demonstrated the efficiency gained by using interchangeable parts to make exact copies.

Blanc's mass production of guns and Breguet's bespoke crafting of pocket watches were very different ways of making things, but both would change our world profoundly. Breguet's watches paved the way for the accurate timepieces that we use now to run our lives and our economies, and Blanc's gun-assembly technique pointed the way to modern mass production. However, at the time, Blanc's manufacturing innovation was regarded as a threat to the established French social and political order; the powerful oligopoly of artisan gun-makers, threatened by the idea that a more automated process was eating into their market, lobbied the government to shut down the mass production, and they did just that in 1807. As a result, just as France was poised on the brink of industrial revolution, the government swept away its foundation.

Although Blanc's methods were repressed in France, his revolutionary idea was too potent to ignore – he had shown that applying engineering standards and systems to manufacturing could lower the skill level required to make valuable products. Giving each worker a defined task increased productivity, and therefore output. This is what Adam Smith had observed of pin-makers in *The Wealth of Nations*[4] twenty years earlier, when he described a visit to a small factory manned by ten people, who together produced 48,000 pins each day. If each worker worked alone, making one pin at a time, the factory would struggle to make two hundred pins each day. Smith realised that the division of labour could be transformative, describing it as the source of 'that *universal opulence* which extends itself to the lowest ranks of the people'. As this way of organising took hold, the variety and the quality of manufactured products increased rapidly, at ever reducing prices, and became a source of increasing prosperity.

Cambridge University economist Diane Coyle describes what she calls 'The Hockey Stick Graph' of prosperity, which 'tootles along, doing nothing very much for hundreds years and then, in the nineteenth century, turns a corner and becomes exponential'. Indeed, since Smith's time, the total real output of the world's economy has soared, from the equivalent of one trillion dollars to 110 trillion dollars,[5] and GDP per capita has increased more than ten-fold. During the same period, some countries have enjoyed even more spectacular growth; the US, for example, has experienced a near thirty-fold increase in per capita GDP. This surging economic growth, which continues today in most countries, has made progress possible in many different areas, not least by enabling the provision of such basic necessities as health care, energy, water and food, and the consequent reduction in the number of people living in extreme poverty. It was triggered by what Coyle describes as the 'fundamental drivers of growth': innovation, the division of labour and exchange. All of these factors started to gather momentum during the Industrial Revolution, allowing people to work more productively and make quality goods at scale.

As a result, many complex manufactured products that started life as expensive luxuries have become mass-produced, and hence affordable. This is one of the most transformative powers of engineering.

EXTREME POVERTY IN THE WORLD POPULATION

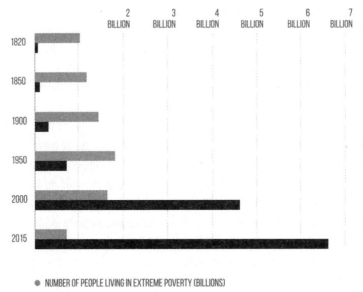

● NUMBER OF PEOPLE LIVING IN EXTREME POVERTY (BILLIONS)
● NUMBER OF PEOPLE NOT IN EXTREME POVERTY (BILLIONS)

FIGURE 2.3 Good progress for people: the decline of extreme poverty since the 1950s.

For a few dollars you can now buy a watch that is just as accurate as a Breguet masterpiece. The mobile phone also rapidly became affordable. There was only one available in 1973; it cost $4,000 and the battery lasted for half an hour. Now there are more mobile phone contracts than there are people on the planet[6] and a well-equipped handset is cheaper than a meal in a mid-priced restaurant. John Hennessy, Executive Chairman of Alphabet and winner of the 2018 Turing Award, the most prestigious prize in computer science, believes that 'engineers have a really attractive feature [which is that] they care about the quality of a solution and the difference it makes'. The ability to make useful things at scale, without sacrificing quality, often makes that difference.

Thomas Jefferson was the United States Ambassador to France during the 1780s, where he saw Blanc's work and quickly grasped its potential. As a result, it was the US that first standardised production practices, eventually mass-producing everything from clocks to bicycles and automobiles. In 1854, the British industrialist Joseph Whitworth wrote glowingly about the 'American System of Manufacturing', saying that 'Wherever [mechanical automation] can be applied as a substitute for manual labour, it is universally and willingly resorted to.' For Whitworth, it was mechanisation, 'under the guidance of superior education and intelligence', that was driving 'the remarkable prosperity of the United States'.[7] At the start of the twentieth century, Henry Ford further improved mass production by breaking down the process of building an automobile into its simplest tasks and wherever possible using tools, rather than humans, to complete them. Many workers were still required, but Ford employed them in a new way, learning from Frederick Taylor's and Frank and Lillian Gilbreth's 'scientific management' systems.[8] They were timed and filmed as they did their jobs and, from these observations, Ford formulated the rules and physical movements that workers needed to follow to complete their tasks. Men and women became replaceable components in a greater manufacturing machine. At his Highland Park production line, the time it took to build an automobile plummeted from over twelve hours to just ninety-three minutes, and a brand new one was completed every three minutes.[9] In the five years from 1909 to 1914, the price of the Ford Model T was cut in half. Automobiles were no longer curiosities for the wealthy few; they were a convenient form of transport for the many.

It was only a matter of time before automobile manufacturers sought even more efficient and reliable production processes. In 1961, General Motors introduced the world's first industrial robot at their factory in New Jersey. Using instructions stored in its drum memory, the Unimate robot could grip, weld, drill or spray, handling loads of up to 500 pounds. All too aware of the possibility of a backlash from workers who might worry that robots would take the best manufacturing jobs, the Unimate's manufacturers were careful to describe their invention as only handling 'dull, difficult or dangerous jobs'. The first machines took on the hazardous task of lifting hot, die-cast car

FIGURE 2.4 Mass production perfected to a T, driving away exclusivity:
Ford's Model T (1910s).

parts from an assembly line and welding them onto an automobile
body. By 1969, a plant in Ohio, making extensive use of these robots,
was able to set a new record, finishing 110 new automobiles every
hour.[10]

THE RISE OF THE ROBOTIC MAKERS

One revolutionary consequence of advances in robotics and
computer-controlled manufacturing was the ability to make things
with an unprecedented degree of precision. The 'Six Sigma' process
was developed in 1986 by engineers at the telecommunications manufacturer Motorola, as a way of reducing deviations in manufactured
products from their designs and thereby improving their quality
and replicability. This process can provide a statistical guarantee
that 99.99966 per cent of parts produced are free of defects and is

recognised as a hallmark of excellence. Silicon microprocessors or 'chips' are the most complex man-made objects ever conceived. For them to work, billions of microscopic components must be built *in situ*, on a wafer of flawless silicon. The Six Sigma standard permits 3.4 defective components per million – but even this is unacceptably high, since any minute error in positioning, connectivity or material purity can render a chip inoperable.

When I joined the board of leading microchip manufacturer Intel in 1997, I was eager to visit a fabrication plant (or 'fab') to see chips being made. Before they let me in, I had to don a hooded 'bunny suit' and be blasted by the powerful jets of an air shower, to ensure I did not carry any lint into the ultra-clean facility; the air inside the fab was thousands of times cleaner than in any hospital operating theatre. I watched the graceful, choreographed movements of robotic machines, as they passed highly polished discs of pure silicon between them, completing every procedure with an astonishing level of fidelity. Things had moved on a long way since my first encounters with the makers of microchips in California in the 1970s, where I would sometimes visit a computer centre to use a computer-controlled pen plotter. I was there to make maps of the sub-surface of the Prudhoe Bay oilfield in Alaska. Most of the other people I met there were drawing computer-generated logic diagrams for printed microelectronic circuits. The detailed circuit diagrams were drawn on large sheets of paper, which were then reduced to millimetre scale and used to make 'masks' to control the lithographic etching of silicon discs.

Semiconductor manufacturers like Intel learned quickly that complete control of every stage of the manufacturing process was critical to their success, because minute variations could have dramatic consequences. Andy Grove, CEO of Intel during its most rapid period of growth in the 1980s and 1990s, introduced the 'copy exactly' manufacturing philosophy. After a production process is perfected in the research and development facilities, it is precisely replicated on a larger scale in a new fab. Everything from the paint on the walls and the quality of light to the colour of the technicians' gloves is

FIGURE 2.5 Dust-free wafers and chips anyone? Fabulous purity and fidelity in Intel's fab.

copied to eliminate any possibility of introducing errors. During the two decades since my first visit to a fab, the electronic components on a chip have continued to get smaller and more densely packed. As a result, electrostatic and quantum forces can create increasingly unpredictable behaviour, providing further potential sources of manufacturing error. Overcoming these difficulties to get a high yield of functioning chips, without increasing their cost, relies on the simultaneous mastery of physics, chemistry and materials science and is one of the greatest successes of modern engineering – the chips made in a fab have changed every aspect of modern life.

With advanced robotics, powerful computers and the ability to 'copy exactly', the scene appeared to be set for automation and robots to dominate the mass production of automobiles and many other goods. However, five decades on from the Unimate, this has not happened. 'You might be surprised,' says Dieter Zetsche, the Chairman of Daimler AG and Head of Mercedes-Benz Cars, 'but our objective is not to accomplish the maximum degree of automation.'

In fact, he explains, there are situations where his company is moving in the opposite direction, and replacing robots with people. Zetsche describes how, previously, his company had built complex assembly lines, where each step was rigidly defined and seamlessly connected to the next. This achieved the aim of maximising productivity, but only for the high-volume production of a limited product range, since it 'created an extremely high level of inflexibility, and [required] high investment if any kind of change was needed'. In today's competitive automobile market, manufacturers need to offer a huge variety of models: gasoline, diesel and electric, as well as a wide variety of hybrids. Consumers also want to have various features customised to create the car that they want. 'When you multiply the different variations available, you come to almost millions [of options],' says Zetsche. In this context, flexibility is extremely valuable, which is also where humans excel. 'Robots have their skills, and humans have their skills,' says Zetsche, diplomatically. He has identified a great opportunity for much better collaboration between the two. 'It's teamwork now. This is the way we are going.' In their production facilities in Germany, Mercedes-Benz have put this powerful logic into action. Zetsche explains that, for safety reasons, old-fashioned industrial robots used to be caged, but the new generation of robot is smaller, more adaptable and, most obviously, uncaged. Arrays of cameras and sensors give them an awareness of their environment and allow them to better communicate with the people and other robots around them. Rather than the automobile creeping inexorably along a rigid assembly line, small autonomous wheeled robots bring components to and from the partly built chassis.

To install a 'heads-up display' that projects driving information onto the windshield, a technician climbs into a nearly completed automobile. A small robot passes him the parts he needs and the augmented reality glasses that he wears allow him to quickly adjust the unit to the perfect angle. This job used to be done by a much larger robot working alone, and it went wrong if either the automobile or the robot were fractionally misaligned. By weaving together artificial intelligence and data analytics, and by harnessing the potential of a

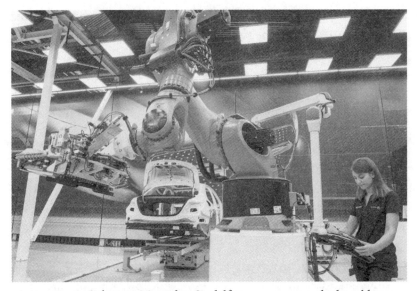

FIGURE 2.6 Robots at Mercedes, Sindelfingen – smart and adaptable to work with humans.

human-machine system, Zetsche believes that his company is effecting a complete re-engineering of the assembly line. 'I think it's fair to talk about a revolution,' he says. 'We now have the most efficient investment, highest flexibility and the highest productivity output ever.'

ENGINEERING LIFE: A NEW FRONTIER

The recent changing face of manufacturing is not just a story of large-scale mechanisation and digitisation. Since Neolithic times, humans have used living cells to ferment wine and leaven bread, and synthetic biologists are now modifying and using cells in much more powerful ways. At Imperial College in London, Paul Freemont, founder of the Centre for Synthetic Biology, describes the potential of synthetic biology with the zeal usually demonstrated by Silicon Valley's technology evangelists. 'There will be, in my opinion, no technological limit,' he asserts. 'Only utility and societal acceptability will hold this technology back.' Genetic engineering has been around since the 1970s – it is what is used to produce the insulin injected by most diabetics, and to generate a small number of other successful pharmaceuticals and genetically modified crop strains, for example[11] – but according to Freemont, 'there is no real engineering in the traditional approach to genetic engineering'. As a result, most genetic engineering applications have been so bespoke and used in such small quantities that they have failed commercially. Biology presents a significant engineering challenge, since living systems are dynamic, non-linear, evolving and replicating, and the conditions inside any single living cell can change dramatically at different times. A genetic engineering procedure that works, for example, in a skin cell may not work in one from a lung; it is therefore very difficult to create a 'copy exactly' process. Freemont recalls how insights from construction and process engineering made him change his approach to biology. He 'saw a completely different way of thinking and problem-solving. I realised that I could do something useful with my biological knowledge, not simply solve fundamental problems to understand how molecular biology works.' Since that moment, Freemont has contributed to the creation of the field of synthetic biology. Synthetic biologists have invented a range of tools that allow them to compose and edit genetic sequences with a new-found precision, reproducibility and reliability. They have adopted the tools of process engineering to introduce rigorous standards and protocols that make the results of their actions more predictable. Biology is becoming an established part of engineering and, if this rapid progress continues and synthetic biology succeeds in unravelling the complexity of cellular metabolism, the implications could be profound. As Freemont explains, 'Evolution is just one end point of billions of years of trial and error;' the tools of synthetic biology would allow bioengineers to rapidly design and test many different adjustments to an organism's genetic instructions, effectively accelerating and directing the course of evolution.

The first products of synthetic biology are now coming onto the market and include a mountaineering jacket, a tie and an ultra-light running shoe that have all been woven from spider silk. Unlike the spider-silk stockings given to the French King Louis XIV in 1709,[12] which were made by meticulously harvesting silk from hundreds of individual spiders, no spiders were involved in the production of these new garments. Instead, this silk was produced by bacteria or yeast whose genomes had been re-engineered to make them produce silk proteins on an enormous scale. This has many more important applications, since spider silk's strength-to-weight ratio and elasticity cannot be matched by any other man-made material; if each strand in a spider's web was scaled up to just a millimetre in diameter, it would be strong enough to stop a speeding train.[13] Many other new and innovative products are also on the way. Biologists believe that it will soon be common practice to create reprogrammed immune cells that seek out and destroy cancers, and there has been technical success in re-engineering algae so that it can convert sunlight into energy-dense liquid fuel.[14] Synthetic bacteria can also produce a range of high-performance bioplastics,[15] which cannot yet compete with the cost of their fossil fuel-derived equivalents,[16] but have the big advantage of not adding to anthropogenic carbon emissions, a factor that is becoming hugely important.

Society needs to be ready to accept the new creations of synthetic biology. This acceptance cannot be taken for granted and recent history shows that rational arguments do not necessarily win the day. When genetically modified (GM) crops were introduced at the end of the last century, they were met with widespread hostility. Many believed that they and the farming practices that they encouraged were unsafe and would severely affect our health and damage our natural ecosystems. Although those concerns were not supported by evidence, they were nonetheless passionately held. By dismissing people's heartfelt concerns out of hand, the scientists and businesses that developed these crops stoked further outrage. Their failure to engage in a more open and constructive dialogue fuelled concern and made the licensing of GM crops a risky political issue in many parts of the world, including the European Union, which now has the most restrictive licensing regulations on the planet. Future success in this area will require engagement on the basis of society's concerns and not on the basis of business's concerns or of scientists' views of the available evidence. If this does not happen, advances in synthetic biology will create greater antagonism and slow the development of the technology and the delivery of its benefits to health and well-being.

An additional factor in all of these debates is that potent pathogens could be made using synthetic biology, for biological warfare. This requires an extraordinarily high level of technical knowhow, a barrier that, for the time being, should preclude rogue actors from manufacturing them. However, biologists like Freemont will need to work hard to show that they can maintain control over their creations. Safeguards will be needed to prevent the creation of dangerous new life forms and diseases, and the new generation of biologists will need to be educated to observe high ethical and practical standards,[17] in order to convince society of the long-term safety of their work.

PRINTING THE FUTURE

At a BAE Systems manufacturing site in the north of England, there is another new way of making objects. You can watch through a window as a laser beam traces an intricate shape on a perfectly flat layer of white titanium powder, sending up small puffs of smoke. The intense heat of the laser melts the powder, laying down a thin layer of solid metal. As the process is repeated, successive layers of titanium are built up. Slowly, a complex, 3D-printed structure takes shape. Engineer Greg Flanagan displays several aeroplane components that used to be assembled by painstakingly welding multiple components together – these intricate shapes, with complex interior structures, are now printed in a single operation. As Flanagan explains, 'one of the beauties of 3D printing is that complexity is free', which means that it opens up huge possibilities with imagination the only limit. It costs far more to fabricate a complex structure by more traditional means, regardless of whether it is moulded, machined or assembled. Another advantage of 3D printing is its ability to create physical prototypes rapidly and cheaply, even from a high-performance material like titanium. This removes barriers to innovation, since makers now have a new freedom to experiment, embedding improvements and adapting to changing requirements faster than ever before. The sportswear manufacturer Adidas recently opened fully automated 'lights out' factories in Germany and in the US. In these factories, each production step is executed solely by 3D printing and robotics – the only task that cannot be automated is lacing the shoes that emerge from the end of the production line.[18] Other exciting applications are occurring in medicine. Biomedical engineers can now, for example, 3D-print replacement knee joints, heart valves and skin, and these bionic parts can be rapidly tailored to fit the precise shape and needs of an individual's body.

And this is just the beginning. The 3D printers that I watched producing parts for airplanes worked precisely but slowly, but they are going to become much faster and more powerful. At present they can only use a small number of metals, resins and plastics, and they cannot switch materials midway through a building process, but that will also change. For example, researchers are already developing ways to 3D-print complex microelectronic and microfluidic circuits.[19] Analysts have projected that 3D printing could eliminate one-quarter of world trade by 2060,[20] since countries would be able to produce what they need at home without the need to import parts or finished goods; that will create both opportunities and challenges.[21] As Kasper Rørsted, the CEO of Adidas, points out, his company's fully automated factories are still the exception. They produce a very small fraction of their total shoe production and, he says, 'it's a complete illusion to believe that manufacturing can go back to Europe [a high cost location] in terms of volume.'[22]

During the 1980s, Eric Drexler, a Senior Research Fellow at Oxford University's Martin School, helped to launch the field of nanotechnology. His dream is to use it to take the principles of the 3D printer to their extreme – he argues that it is possible to build machines that would manufacture products by bonding individual molecules, in precise positions, to a growing structure. Feed one of these hypothetical machines some simple chemical building blocks, input any

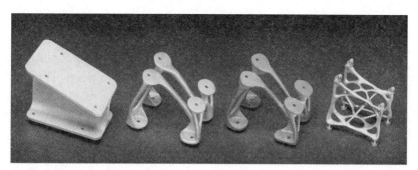

FIGURE 2.7 3D-printing adds a new dimension to manufacturing: airplane cockpit brackets (traditionally manufactured, *left*; 3D-printed, *first three from right*).

design, and the machine would make it – in principle, there would be no limit to the structural complexity of objects that could be produced. Anything from advanced computer microchips to multi-layered solar cells and complex catalysts that would disassemble carbon dioxide could be synthesised rapidly. The late Richard Smalley, a Nobel Prize-winning nanotechnologist, dismissed these ideas as wishful thinking. He argued that, as a chemical substance grows, the successive addition of each atom or molecule depends on the position of all the other nearby atoms, and asserted that this intrinsic interconnectedness between atoms would make the reactions impossible to control. Drexler, however, says that, while there are many technical hurdles to clear before his idea could be delivered, it is no idle fantasy. With some irony, he tells me, 'I tend to avoid speculation. The work that I have been doing is such that, if the analysis is wrong, one can almost say the physics must be wrong, the textbooks must be wrong.' So far no one, including the US government, has committed to taking this further.[23] I think there is a high probability that his ideas will never work.

Although we have not yet mastered the ability to build at the level of individual atoms, engineering at the nanometre scale is already changing many facets of our world. It is a challenge to merely visualise objects that exist at this scale, let alone to conceive of ways to manipulate them in a precise and controllable way. To give some perspective, the hairs on the back of your hand, which are among the finest structures we can see with the naked eye, are about twenty thousand nanometres wide, which is immense to a nanotechnologist. Battery manufacturers are improving the performance of their cells by engineering their electrodes at the nanoscale. Most conventional lithium ion batteries use graphite anodes, but it has long been known that silicon anodes can store up to ten times more energy. The problem is that they break down rapidly, but by controlling their structure at the nanoscale, engineers are now developing lithium ion batteries with much more stable silicon anodes and greatly improved energy density.[24] A growing number of pharmaceuticals are also based on nanotechnology. Molecular patterns engineered on the surface of drug delivery capsules can, for example, ensure that drugs are delivered specifically to infected or damaged cells. Today's most powerful chips also depend upon nanoscale engineering; the transistors are just ten nanometres wide, and getting even smaller.[25]

MAKING AN AUTOMATED FUTURE THAT WORKS FOR ALL

Manufacturing processes are becoming more and more automated, reviving old fears that they will cause widespread unemployment, inequality and social upheaval. Ever since the Industrial Revolution in Britain, those fears have returned again and again. One of their first manifestations was in the early 1800s, when the Luddites went on the rampage in England, smashing the knitting frames and power looms that they believed were stealing their livelihoods. As this was going on, the economist David Ricardo warned that 'the substitution of machinery for human labour ... may render the population redundant'.[26] I asked my friend, the Nobel laureate economist Mike Spence, for his view on the current wave of fears and if he thought new technologies would fill the world with unemployed people. Relaxed and expansive, Spence has a gift for clarifying the knottiest of economic scenarios. 'Technology all along has produced changes of structure and disruption in the labour market, and this seems to be no exception to that,' he began, injecting a measure of perspective.

'And, when you're in the middle of a difficult transition, it's easy to think that something has gone desperately wrong'. During past labour market transitions, forecasts of massive unemployment have not materialised. New technologies replace *tasks*, but not necessarily *jobs*. In fact, by automating defined tasks, new technologies often *create* work, rather than destroying it. The kind of machines that the Luddites raged against did precisely that. Automation removed 98 per cent of the manual labour needed to weave each yard of cloth, but weaving employment did not shrink by 98 per cent, because the falling price of woven cloth stoked demand. Weavers adapted to this new world, and learned how to supervise the new machines. As a result, the number of weavers actually increased – in the US there were four times as many of them in 1900 than there were in 1830.[27]

FIGURE 2.8 LOL: the leader of the Luddites stoked fear and anger among workers about being replaced by machines (1812).

New technologies create new industries. Six decades ago, there was no global software industry; now it is a US $300 billion business, employing 20 million active developers.[28] Much of today's most important making goes on in this digital space, and some of today's most important tools are blocks of code. Overall, while computers and automation have made certain tasks and roles obsolete, they have not caused any net increase in unemployment.[29] Of the 271 jobs listed on the 1950 US census, for example, there is clear evidence that only one of them, elevator operator, has been eliminated by automation.[30] Instead, as industrialised economies mature and wealth percolates through society, the demand for high-quality goods and services and better infrastructure grows. E-commerce has automated many tasks in retail but it has also fuelled retail spending, creating more than 300,000 jobs in the US since 2007. Experts have repeatedly failed when trying to predict the type and scale of new jobs. Just two decades ago, who would have confidently forecast the need for thousands of social media analysts, search engine optimisers and smartphone application developers?

As well as creating work, automation tends to improve working conditions. Mike Spence reminds me that, on average, working conditions get safer, cleaner and less tiring as industrial economies mature. In the US, the number of accidental deaths at work dropped dramatically during the twentieth century, from more than sixty deaths per 100,000 workers, to fewer than five.[31] The working week fell from an uncomfortable global average of over sixty-four hours in 1870, to thirty-six hours in 2000.[32] In 1930, the economist John Maynard Keynes predicted that technology would increase productivity so much that, by the end of the twentieth century, nobody would need work more than fifteen hours a week.[33] Things have not changed so quickly, but Spence thinks that the trend towards a shorter working week should continue – 'it may be that we are cheerfully, happily working twenty hours a week', he says. At the same time that automation has created employment, people have demanded a more equal balance between work and other aspects of their lives. Access to mobile computing and changes in labour contracts are entirely blurring the boundary between work and home life; the trend is to 'work-life integration'. Many of the latest technological innovations, such as social media applications, are aids not to productivity but to leisure and means of entertainment. Might this be one of the clearest signs that progress is real? Since many people spend less time earning a living, they can devote more time and energy to their leisure, as this figure for the industrialised world shows.

The economist Robert Gordon paints a gloomy view of the future, in which the heady days of rampant economic growth are gone. He attributes growth to the technologies that societies make, and argues that the one-hundred-year period from 1870 to 1970 delivered innovations that radically transformed the world for the better.[34] The internal combustion engine let us move freely. Indoor plumbing removed squalor and sickness. Refrigerators and air-conditioners kept food fresh and hot summers productive. Penicillin and chemical fertilisers kept us alive and well-fed. These, and many other modern marvels, appeared and flourished during these ten decades, creating a once-in-history level of economic growth and a consequent improvement in the average

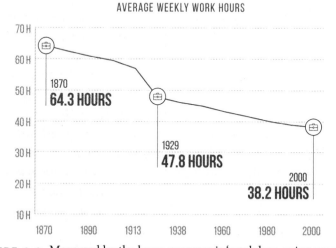

FIGURE 2.9 Measured by the hour: progress is 'work less, enjoy more'.

quality of life. There was a huge improvement in people's productivity, new opportunities for businesses were created and bigger wages were paid to workers. In contrast, increases in productivity since 1970 have been sluggish, and wages in the US have grown more slowly. Gordon argues that, while contemporary technology such as smartphones might be impressive and make daily life a bit more convenient, compared to what came before they represent minor, incremental steps forward, which have consequently made minor, incremental contributions to economic productivity.

I agree with both Spence and Coyle, neither of whom is convinced that Gordon's pessimism is well founded. Instead of hailing the end of economic growth, Spence suspects that the advanced economies that have embraced digital tools are poised for another uptick in productivity. 'We've seen this movie before,' he says, pointing to a mistimed quip made by a fellow Nobel Prize-winning economist. In 1987 Robert Solow wrote, 'You can see the computer age everywhere but in the productivity statistics.'[35] He was too hasty to suggest a 'computer productivity paradox', since productivity did eventually rise, but only in the late 1990s – which demonstrated that there is always a lag between an innovation and its impact. During the last two decades of the nineteenth century, for example, electric power stations were springing up in major industrial cities all around the world. Coyle points out that there was, however, no sign that electricity, the most eagerly promoted technological advance of the age, was improving productivity. Factories had been built to run on steam power – their activities tended to be arranged over several floors, clustered around a single large steam engine. At first, the central steam engine was simply replaced with an electric dynamo – it ran more efficiently, but little else changed. Only later, during the early twentieth century, did industrialists realise that it would make more sense to reorganise factories completely, arranging them on a single floor and equipping each machine with its own electric motor, which operators could control independently.[36] These new factories were much more productive, though building them took a long time and cost a lot of money.

There is a lot of talk about the immediate impact of 'exponential technologies', but most transitions in manufacturing still unfold at a relatively slow rate. 'The industrial manufacturing sector moves so slowly. It moves slowly in every dimension,' says Professor Andy Hopper, Head of Cambridge University's Computing Laboratory. He, like me, has seen how long it takes to integrate new digital manufacturing technology into the industrial sector. Investors are reluctant to take a risk with new equipment if it imperils their output.[37] Most believe that there is a 'first mover disadvantage', which is gradually removed over time, since technology gets more reliable and cheaper – and therefore waiting is often a winning strategy.[38]

Coyle offers another note of caution for anyone trying to analyse current economic trends. Conventional economic measures of productivity, such as GDP, do not tell you much about the many dimensions that contribute to quality of life. 'They take highly uncertain statistics and divide them into each other or partition them in moderately arbitrary ways,' she explains. The digital revolution is delivering a rush of new services and capabilities, many of which are free. Some of these can profoundly change the way we live and work, but measuring the true value of the free encyclopedia Wikipedia, free communications like WhatsApp and free navigational tools like Waze is very challenging. That is also true of innovations in other sectors, such as health care. 'We've seen a startling increase in longevity, and I would say it is very difficult to know to what extent we've measured the impact of any of that economically,' says Spence. Traditional measures of productivity were designed for a time when economies were dominated by manufacturing, but they are no longer fit for purpose.

Coyle knows from personal experience that the automation of manufacturing can create turmoil on a local level, even if it is not reflected in national statistics. She grew up in an industrial town in the north of England, and her father lost his job as a result of automation in the early 1980s. However, she lays blame not on that but on the designers of social policy. 'We know that this is what automation does. We didn't have an adequate policy response when my father lost his job. So why do we still not have one now?' she asks. For her, the cornerstone of society's response should be obvious: 'Education, education, education', she exclaims. Coyle is not convinced that wealth redistribution schemes, such as the widely publicised 'universal basic income', will ever be a long-term solution, describing them as 'an individual solution to a collective problem'. She points out that giving somebody a small amount of income does not help them either find a new job, travel to the new job, get good health care or pay for the training they need to apply for a new job. Instead, Coyle prefers to think about 'universal basic infrastructure'. High-quality schools and colleges would be a crucial pillar of that infrastructure. Better access to good education, and to the re-education needed for workers to keep up with changing technology would also help tackle the so-called 'skills gap', which opens up when new job opportunities, such as writing code or maintaining automated systems, demand sophisticated technical abilities and experience. The risk is that displaced workers who cannot obtain these skills find themselves competing for low-wage, low-skill jobs, while employers who need more qualified employees are starved of talent. 'I won't say that re-education is the whole answer,' says Spence, 'but it's certainly an important part of the answer. It deals with employability, it deals with contributing socially and it deals with a lot of things that income redistribution by itself does not.'

MASS CUSTOMISATION: EFFICIENT, PERSONALISED AND ON-DEMAND

For two centuries, the manufacturing sector has been the engine of global economic progress. In both developed and developing countries, manufacturing contributes disproportionately to exports, investment in innovation and productivity growth,[39] but the number of people employed directly in it is likely to diminish. Work is not drying up, but the nature of what people do is changing.

At the beginning of this chapter I introduced two French engineers: Blanc, the mass-producer, and Breguet, the master craftsman.

In the eighteenth century, they occupied opposite ends of the manufacturing spectrum; had they been alive today, their work would have been much more closely aligned. To make their bespoke watches, Breguet and his assistants had to make many of the components by hand. Today, bespoke makers have access to huge catalogues of standardised parts, an idea pioneered by the industrial manufacturer Blanc. I saw the power of this approach when I visited Surrey Satellite Technology. Its CEO, Martin Sweeting, explains that 'the first four decades of the space era were dominated by a few superpowers who alone possessed the knowledge and budgets to undertake the enormous technical and programmatic challenges posed.'[40] That has changed dramatically during this century. In a large room, Sweeting shows me a range of micro-satellites under construction. These have altered the economics of space profoundly; the key to reducing costs has been the enormous advances in the manufacture of microelectronics. Most of these small satellites are made from off-the-shelf components that are also used in consumer electronic devices. Some of them use a smartphone to navigate and take images of Earth.[41] 'Space is now within the reach of small companies, universities and even high schools,' says Sweeting. The falling cost and improving capabilities of constellations of small satellites will stimulate new applications in communications, Earth observation and more.

The makers of small satellites solved the challenge of integrating and adapting existing technologies to create new and powerful capabilities, bridging the difference between the bespoke and the mass-produced. Nanotechnology and synthetic biology promise to bridge this same gap by, for example, creating materials and cells that can self-heal and adapt themselves to changing environments. Dave Holmes, the Manufacturing Director of BAE Systems, anticipates a new generation of factories that will combine disparate automation technologies, including 3D printing and robotics, into systems that can rapidly be reconfigured for every new product line, without the need for any retooling or refitting. It will make little difference whether an order is a one-off or a run of thousands. For Holmes,

FIGURE 2.10 No-frills space travel: low-cost satellites are the way.

these factories of the future are 'well within the realms of what we'll see in the next decade'.

The team of young makers in a London start-up called Unmade provide another glimpse of the possibilities created by this new world of mass-customisation. In a studio beneath London's Somerset House, a centre for art and culture, Ben Alun-Jones points to one of the grey automatic knitting machines that is controlled by his company's software. It is the length of a small automobile, with spools of brightly coloured yarn lined up along the top. 'That one probably has 30,000 moving parts. It's far more complex than any 3D printer,' says Alun-Jones. He describes how the machine is entirely computer-controlled, although the proprietary software that comes with it is frustratingly inflexible. Alun-Jones and his co-founders have set about adjusting the machine's code to develop new capabilities. Some of the results are displayed on a nearby rack that holds an eclectic mix of colourful knitted sweaters. One bears a four-colour portrait of Ernest Hemingway, while another disturbs the neat blue stripes of a classic Breton sweater with a chaotic series of swirls and vortices. Unmade's new code gives customers a role in the design of their clothes and, at almost no cost to efficiency, every garment made by the machines that are controlled by the company's software is unique. 'You could do what we're doing by manually programming each one. We do it on an automatic industrial scale now,' says Alun-Jones, who hopes that this new approach will soon expand to tailoring. 'Fit is much harder than what we're doing at the minute ... But it's on our road map. When we solve that, it will be a big deal.'

Alun-Jones is also committed to helping to reduce the extraordinary waste created by today's fashion industry. 'It's a $2.4 trillion industry and there is $250 billion a year in wasted stock,' he explains incredulously. Both Unmade and Mercedes show us that manufacturing will increasingly shift to an 'on-demand' process, in which products are tailored to individual requirements and made only when they are required, in a drive to minimise waste and maximise efficiency.

The same drive has also resulted in aluminium drinks containers being around 40 per cent lighter than they were in 1980[42] and lightweight construction materials being used in automobiles and airplanes to increase their fuel efficiency. It has also given birth to sensors and computer systems that give consumers and producers information to allow them to reuse and recycle materials in more efficient ways. The shipping company Maersk Line, for example, introduced a 'Cradle to Cradle Passport', which will track the components of vessels throughout their life cycle, significantly increasing the amount of material that can be recycled when a ship is broken up. And rather than recycling paper, an Israeli firm has created a reuseable paper system which, when it erases the ink, digitises the information on a page and stores it in the cloud.[43] With advances in synthetic biology, plastic landfill and other waste streams can increasingly be converted into valuable sources of raw material and energy.[44] These approaches to eliminating waste have not only been good for the environment but have also improved the reputation and long-term profitability of the companies involved; more could be done to make these approaches a standard part of manufacturing.

Innovation always builds upon what came before and bears the imprint of many hands and minds. Tony Fadell, one of the principal inventors of the iPod and the iPhone, argues that this is why the open-source movement in software and hardware development is so revolutionary, and why it has become a standard part of making things. He explains that, before open-source, most large technology corporations kept their source codes and hardware designs to themselves, which 'did not allow for evolution to happen all around the world'. Then, as Fadell says, 'we unlocked it with open-source'. Open-source Linux software keeps the world running, as it controls most servers and forms the basis of both the Google Android operating system and the systems that control modern automobiles. The tremendous uptake of these systems is a powerful illustration of what can be achieved when innovators are allowed to pool their expertise and create tools for common use.

Fadell thinks the future potential of the open-source approach is most vividly illustrated by what is happening in China's high-tech hubs in Shenzhen and elsewhere. 'I am incredibly in awe and a bit scared about what's going on there,' he says of the blistering pace with which new code and new devices are being developed. He describes the way that many developers there have a completely different approach to intellectual property and collaboration. 'What we would call "stealing", they might call "sharing",' he says. The fact is that much can be shared but not everything is open – however, if you have been brought up in the era of open-source, you may think that everything should be open. Fadell echoes his former collaborator Steve Jobs in quoting Picasso's famous line that while 'good artists copy, great artists steal' when comparing developments in China to the US at the turn of the twentieth century. Nigel Whitehead, Chief Technology Officer of BAE Systems, says that during last two decades 'there's been the greatest migration of intellectual property in the history of business worldwide, which has been stolen out of the computers ... of

Western businesses [by computer hackers]'. That should not become a standard practice, and I will talk more about it in Chapter 8.

PHYSICAL THINGS IN A DIGITAL WORLD

In 2011 the web pioneer Marc Andreessen announced that 'software is eating the world'.[45] His argument was that every part of our lives and every corner of our economies is being reduced to bits and bytes, which are then manipulated by computer algorithms. The implication was that the physical world, populated by the tangible objects that we make, build and exchange, will gradually recede in importance. While it is true that inventions such as the smartphone collapse multiple functions into a single form (nobody need carry a separate camera, diary, typewriter, pocket radio and book with them these days), these positive trends towards 'dematerialisation' and 'digitisation' can only go so far. As John Hennessy puts it, 'If we're really going to invent a better world ... there's got to be more than just digitisation.' We cannot eat software and nor will it heal us, clothe us or provide us with shelter. We will still need such basic things as steel pipes and optical fibres, as well as the more complex technological devices that support all our digital tools.

Fadell is adamant that innovation in hardware still represents the way forward. 'Software can only eat the world after a hardware disruption has been created. If the hardware wasn't there, none of the software would have been created.' And in his view, getting the devices right so they can unshackle new potential is critically important, and often much more challenging than software engineering. 'It takes a radical way of thinking about the entire system to create the explosion of new opportunities, new businesses, new industries and new verticals that can be created [after a transformative new device or platform is invented].' The successful launch of constellations of GPS satellites, which I will cover in Chapter 7, is a clear illustration of this point – only after the physical satellites were in orbit and fully operational could the crucial applications upon which we all now depend be born. The European Space Agency's Paul Verhoef echoed Fadell's central message when he said that, as hardware innovators, 'we need to look ahead, often a lot further ahead than where the IT folks are looking'.

Many of the biggest challenges facing humankind can only be tackled by making physical objects and systems. The crucial goal of reducing carbon dioxide emissions in order to tackle climate change demands that we innovate and deploy new energy sources and distribution networks. Relieving energy poverty and securing supplies of water, food, sanitation and other basic necessities in the developing world requires physical engineering. As health care improves and populations get older, we will need better and more affordable medical devices that we can deploy ubiquitously. And, as cities grow, we must make physical infrastructure and transport systems that maintain cities as functional and pleasant places to live. Digital tools can certainly help us make solutions in more inventive and efficient ways, but they cannot confront these problems alone.

The engineering of physical objects at higher and higher efficiencies is, however, the primary means through which the fruits of progress can be shared most widely. 3D printing and related fabrication techniques will take this to a new level, allowing for more on-demand, localised and customised manufacturing, with less wastage of material and energy. Despite repeated claims that software will 'eat the world', it is of no value without hardware to run on. The relatively slow pace of hardware development will always limit the rate of change in manufacturing and technological advance more generally, though manufacturing will increasingly be changed by robotics, data analytics, connected sensors, synthetic biology and nanotechnology. Society will use these new tools to continue the decades-long trend of reducing average working hours. If our basic needs can be met with less effort, we will be able to invest more energy in improving the world around us.

The human urge to *make* is at the heart of all progress; it is implicit in every new work of art, every attempt at DIY and every patch that is coded for a piece of open-source software. It is a universal activity and the products of engineering allow us all to tinker, refine and choose how we want to make our lives better and more enjoyable.

3

Think

Shortly before she died, my mother asked me to take her to visit the United States Holocaust Memorial Museum in Washington, DC – she wanted to see if she could find the records of any of her relatives who, unlike her, had not survived the Holocaust. We also went to see the museum exhibition, and as we were walking around it, I was astonished to see a desk-sized metal device, painted in utilitarian black and white, and packed with electrical circuits, slots, dials and precise moving parts. The caption described it as an 'IBM Hollerith D11 automatic tabulator'. It looked so bland and unthreatening – why was this precursor to the modern computer displayed so conspicuously, in this memorial to one of the greatest tragedies of the twentieth century?

The object's label provided the answer. During the Holocaust, the Nazis oppressed and murdered with utterly inhuman efficiency, and automatic tabulators made it possible for them to identify those with Jewish blood with great speed and precision.[1] When German forces occupied Hungary, my mother's country, they discovered data from a detailed census that had been conducted by the authorities there. They converted each person's data into a pattern of punched holes on a small paper card; when these cards were fed into a Hollerith machine, electrical brush sensors detected which holes had been punched, logging millions of results quickly and accurately. This allowed analysts to organise and sort census results with great flexibility, so they could

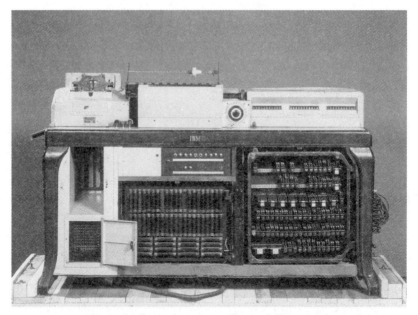

FIGURE 3.1 Misused effectively, a very effective killing machine: the Hollerith tabulator at the Holocaust Memorial Museum, Washington, DC.

pinpoint in seconds all the people with Jewish relatives. Those results led my mother and several hundred thousand others to be packed into freight trains and distributed among camps across Europe, with punched cards also used to schedule and organise these trains with unprecedented punctuality. The people loaded onto the trains were stripped of their names; their humanity was replaced by a number, their identities reduced to a series of holes on a punched card.

The Nazi regime purchased thousands of Hollerith tabulating machines from IBM's German subsidiary.[2] While these machines were not true computers, they were an important step on the long road to today's ubiquitous computing devices. Hollerith machines could sort data, cross-index and perform certain calculations with a speed and precision that the human mind could not remotely match. Since then, devices and systems that complement and extend our powers of thought have changed our world. As the Nazis' eager adoption of the tabulator reveals, people can misuse computing power to cause great harm. However, we can also choose to apply our computer-augmented intelligence to create freedom and opportunity.

These days, we are all so accustomed to machines that help us think that we do not even notice them. Each day, from the moment I get up, my waking life is shaped by calculations. The food on my breakfast table arrives fresh as a result of complex global supply chains. The electrical grid that powers my home seamlessly adapts to the morning surge in demand. A glance at my smartphone brings my emails, as well as news reports and the weather forecast. My car plots the best route to my office through rush-hour traffic. All of these, and so many other features of the modern world, are underpinned by computers. And all this essential, unseen mathematics is executed by swarms of invisible electrons that inhabit the complex labyrinths of computer chips. The engineering accomplishments that make this possible are the most impressive examples of technological progress – and by helping us to think in new and powerful ways, computers have cleared the path for innovations that now affect every aspect of the way in which we live.

We entrust an expanding array of important decision-making processes to computer algorithms. Many of these are based on machine learning, a form of 'artificial intelligence' – rather than operating according to inflexible, rule-based instructions, they instead attempt to learn from previous experiences. The workings of these systems are complex and subject to constant change. Although an increasing amount of their output appears to be the product of some sort of genuine intelligence, we must remember that these systems can still only provide insights based on deductions from past events. And anyone who worries about the arrival of supremely powerful artificially intelligent machines should consider the challenges of maintaining machines even today – the unpredictable behaviour of an office printer is a good example. Computers already outperform human intellect in many specific tasks, and engineers will continue to build and program computers that support us in many ingenious ways, but silicon-based intelligence with the same qualities as human intelligence has yet to arrive.

THE FIRST MECHANICAL AIDS TO THOUGHT

Before the twentieth-century computing revolution, there were very few devices to help make calculations. The abacus was invented in Sumer (modern-day southern Iraq) in the third millennium BC,[3] and allowed traders and mathematicians to speed up their arithmetic. Ingenious astronomical calculators were later devised in ancient Greece – in the second century BC, both the sophisticated clockwork mechanism of the Antikythera mechanism[4] and the simpler astrolabe allowed scholars to keep track of the heavens and make precise predictions about eclipses and other astronomical events. However, these early calculating devices were limited in their scope and utility – they could be used only in very specific operations. In the nineteenth century, things began to change.

In the summer of 1821, the respected astronomer John Herschel paid a visit to his friend, the mathematician Charles Babbage. Herschel carried with him a new set of hand-calculated astronomical tables. Babbage and Herschel scrutinised the figures and found them to be laced with errors and contradictions. Babbage is reported to have said in exasperation, 'I wish to God these calculations had been executed by steam.'[5] He realised that the accuracy of numerical tables was extremely important: engineers needed them to build reliable structures; scientists to make sense of the world; financiers to run economies; and navigators to plot safe passages. Mistakes could have serious consequences.[6] Crucial calculations were made by hand for a long time: the astronomical tables produced in ancient Greece,[7] the ballistics tables used to aim mid-twentieth-century artillery and even the trajectory of early spacecraft were all calculated in this way. The people who did this tedious work were referred to as 'computers', and it is in their honour that we gave today's inanimate machines the same name.

Two centuries ago, Babbage resolved to build a mechanical calculating device that would replace the slow and error-prone human computers. Within a year, he had drawn up the detailed plans for his first 'Difference Engine'. He obtained substantial government funding and recruited a skilled engineer to build it, but it never advanced far beyond its blueprint.[8]

Yet today, in London's Science Museum, stands Babbage's Difference Engine No. 2. Taller than most adults, it is as long as a car and weighs the same as a fully grown elephant. It is powered by a hand crank, which brings a complex sequence of gears, linkages, drives and cams into action. The mechanism literally grinds through complex calculations, and can produce results that are up to thirty-one digits long. The device can even print its results on paper, producing, for example, the accurate tables of logarithmic and trigonometric functions. This difference engine, which was completed in 2000 according to Babbage's plans, took a team of curators and engineers seventeen years to construct, at a cost of several hundred thousand pounds.[9] Importantly, it is a vindication of Babbage's confident assertion that machines, even those based on purely mechanical components, could match or exceed the calculating powers of the human brain.

FIGURE 3.2 Tried but never tested: modern model of the Analytical Engine by Charles Babbage (1834–71).

However, Babbage wanted to go beyond making more reliable tables. As he wrote in his 1864 memoir, 'The whole of arithmetic now appeared within the grasp of mechanism.'[10] In his plans for the Analytical Engine, a more sophisticated calculating device and his focus during the final decades of his life, there is a foretaste of the logic that runs today's computers. Unlike the Difference Engine, the Analytical Engine was programmable. Had it ever been completed, Babbage could have programmed it to operate a wide range of numerical operation by feeding it pieces of card, punched with specific patterns, just like those used by the Nazis to analyse census results and assist with the logistics of the Holocaust.[11]

Despite designing the essential features of a general-purpose, programmable computer,[12] it seems that Babbage may not have fully understood the implications of his life's work. However, one of his close friends would succeed in identifying the full potential of his invention. Ada Lovelace's father was the poet Lord Byron and her mother was a dedicated scholar of mathematics and science. In 1833, she met Babbage at a party in London, and he showed her a functional segment of one of his Engines. She was quick to spot its potential and bold enough to suggest that she work as his assistant. She proved to be a devoted and creative collaborator – in an 1843 letter to Babbage, she wrote, 'I am working very hard for you; like the Devil in fact (which perhaps I am). I think you will be pleased. I have made what appears to me some very important exclusions and improvements.' It seems that Babbage quickly grew to understand and appreciate Lovelace's vision. He wrote back, telling her, 'the more I read your notes the more surprised I am at them and regret not having earlier explored so rich a vein of the noblest metal.'[13] This was unusual in conservative Victorian England, where women were rarely considered men's equals. Lovelace's key role in the history of computation has only recently been sufficiently recognised. The apex of her achievement was her 1843 description of the sequential, algorithmic set of operations required to solve a specific class of equation. To all intents and purposes, this is the world's first computer program,[14] made even more remarkable by the fact that she composed it for a machine that did not even exist. Cambridge University computer scientist Andy Hopper makes the point that, before anyone else, Ada Lovelace 'had the idea of the numbers representing things other than numbers'.

Today's ideas of general-purpose computing are all founded on Lovelace's brilliant work. The numbers manipulated by computers can represent any real-world entity. Lovelace first suggested that they could represent musical notes,[15] but they can also stand in for letters, words, chemical elements, financial transactions, vehicle positions, weather fronts or ethical constructs. It is anachronistic to call Lovelace the first computer programmer, since no computers existed in her time, but the conceptual leap that she made marked the crucial shift from pure calculation to genuine computation. Or, as Hopper puts it, a computer is 'not just a big calculator, it is manipulating information'.

THE FIRST ELECTRONIC COMPUTER

Hopper explains that 'There are many who claim to have built the first computer.' His conclusion is simple and diplomatic: 'They're all true.' One machine with a claim to that title is EDSAC (the Electronic Delay Storage Automatic Computer), which was designed and built by an engineer called Maurice Wilkes, one of the fathers of wartime radar computing. It was first used on 6 May 1949 at Cambridge University. I visited a reconstruction of it at the UK's National Museum of Computing[16] and, even though I have worked with computers for five decades, this one took me by surprise, because it did not look like a computer.

I walked into its innards, where three rows of eight-foot-tall shelving units hold rows of three-inch-long metal cylinders and bulb-like contraptions. These are the vacuum tubes, which replaced Babbage's complex mechanical couplings; they could manage and transform information by switching on and off the flow of electrons.[17] Some of EDSAC's vacuum tubes glow gently, exuding a comforting warmth. There is no monitor in sight, no keyboard and certainly no mouse. Within the tangles of exposed wires, the racks of hand-soldered circuits and the thousands of vacuum tubes is the entire logic of the modern computer.[18] EDSAC is a landmark result of the scramble to develop sophisticated computing machines during the middle decades of the twentieth century.

Urgent necessity drove the world towards the modern computer. By the start of the Second World War, military technology had advanced to such an extent that armies could no longer rely on the instinct and intellect of soldiers and their commanders. For the first time, artillery gunners had to hit aircraft, which were fast-moving targets. Generals disseminated encrypted orders along telegraph wires and across radio waves. Winning the information war would give code-makers and code-breakers an immense advantage. And of course, the war was only brought to an end by the hellish power of the atomic bomb. Building that bomb required complex calculations, many of which could only be executed by computing machines. While the war proved a source of innovation for many advances in computing, most of the solutions invented were tailored to a specific purpose. For example, Hollerith tabulators could only sort and process data that was defined in very rigid ways, while Alan Turing's 'Bombe' and the Colossus computer could only decrypt different types of German messages.[19] At the end of the Second World War, no one had yet succeeded in building the first general-purpose computer.

EDSAC may not have been the first stored-program, electronic, digital computer, but it was the first one that was flexible and practically useful.[20] Though the original machine was only used for a decade before being scrapped, during its short lifetime it contributed directly to four Nobel Prize-winning scientific discoveries.[21] The physiologist Andrew Huxley, for example, used EDSAC to analyse his data, which showed how biological neurons, core components of the original computers, transmit information.

By today's standards, using EDSAC and the other early computers was a visceral and involving experience.[22] The concept of software did not really exist at the time, and Wilkes and his close colleagues had to invent a whole syntax and grammar for programming as they went along. Hopper remembers hearing Wilkes make the pointed remark that 'the older people have not got a clue about what we are doing'. It was that exposure to something new and exciting that made me, at the age of twenty, want to learn how to code, in the old-fashioned scientific programming language of FORTRAN. It struck me as exciting to be able to instruct a machine to do sophisticated things in a logical way. Hopper observes that, even today, coding is seen by many as a mysterious art, practised by young people employed by a small number of technology firms. As we come to rely on software in ever more areas of our lives, that should surely change.

EDSAC, like all digital computers, ultimately dealt in a binary pattern of ON/OFF signals.[23] This was a significant departure from Babbage's engines, which had used the more familiar, but less flexible, decimal (0 to 9) numerical system. The seventeenth-century German mathematician Gottfried Leibniz is usually credited with inventing a system of binary coding, though I believe the germ of this powerful idea developed a generation before Leibniz, in the rich mind of the English polymath and clergyman John Wilkins.

In 1641, Wilkins, one of the founders of the Royal Society, explained how a simple binary code could be used as a universal language to express anything at all.[24] In the convoluted language of the time he said, 'whatever is capable of a competent difference, perceptible to any sense, may be a sufficient means whereby to express the cogitations ... It is sufficient if [these differences] be but twofold, because two alone may be well enough contrived to express all the rest.'[25] This extraordinary fundamental insight would, some three centuries later, launch the information age.

The first programmable computers dramatically reduced the time and effort needed to make binary coding a flexible computational paradigm. Operating the computer was, however, still far from straightforward.[26] Hardware failures were a recurrent frustration during the early decades of computing.[27] As a junior engineer working in Alaska at the beginning of the 1970s, a full twenty years after the development of EDSAC, I worked intensively with a CDC 6600 computer. The programs I ran in FORTRAN to simulate the flow of oil out of the newly discovered Prudhoe Bay oilfield were stored on large stacks of punched cards, and letting a single card get out of order would destroy the whole process. I also learned only too well why we say that a computer has 'crashed'; the 6600 was one of the first systems to rely on a magnetic disk-based memory system, and occasionally the read/write head would literally 'crash' into the disk, wrecking the program run. It was imperative to understand what the computer was doing, and I always checked the calculations on 'the back of an envelope' to confirm that it was actually working as expected. Despite these frustrations and setbacks, it was clear to me that computers were opening up a new world of possibility, and throughout my career they have helped me to think in new ways.

To most modern eyes, the first generation of electronic computers seem intimidating, ugly and somehow alien. But for me, the most surprising thing about EDSAC is how few of the fundamentals have changed in seven decades. If it were provided with a much

FIGURE 3.3 The WEIZAC (1955) and me (1948), at the Weizmann Institute, Israel (2018).

bigger memory it could, in principle, execute any of today's most sophisticated algorithms, though it would take an extremely long time, involve thousands of metres of punched paper tape and probably several vacuum tube replacements. In Andy Hopper's opinion, the construction of EDSAC and the other pioneering computers of the time was 'a bigger computational step in one go than ever before or ever since'. 'Yes, it's true that technology moves fast, and so on,' he continues, 'but even in computing, [it's easy to forget that] some of the biggest moves may have been made a long time ago.' By the early 1950s, the key components of the general-purpose computing machine were in place, and computers were about to become smaller, faster and much more powerful.

THE SILICON REVOLUTION

The Intel 4004 microprocessor, built in 1971, was the world's first engineered 'computer on a chip'. To the naked eye, it looks like a small, sixteen-legged crustacean poised to scuttle away. But under a microscope, one can see a wafer of pure silicon, less than a millimetre thick, on which lies an intricate pattern of minute electronic components. Each link in this complex network is just ten microns wide (one hundred of which could line up side-by-side in a one-millimetre gap). All the thousands of transistors, resistors and capacitors needed to store and process data are built into this circuitry, and on one corner of the wafer is etched a small 'FF', the signature of its creator, Federico Faggin. The EDSAC computer had filled a large room, but the Intel 4004 microprocessor packed much more computational power into an object smaller than a fingernail. Not only was it much smaller, it also cost a fraction of the price. Crucially, it was very much faster, more reliable and easier to program than its cumbersome ancestors.

There had been a great deal of technological progress in the computing and electronics industries in the two decades between the first computers and the Intel 4004. Firstly, the transistor had been invented – in December 1947 at Bell Labs, by Walter Brattain, John Bardeen and William Shockley. At the core of their improvised contraption was a crystal of germanium, a semiconductor whose unique electrical properties allowed the transistor to switch currents on and off rapidly, or amplify them. These are the same core properties of the vacuum tube, but transistors consume much less power and are much quicker and more stable. Their introduction launched the modern computer age.

The production of transistors really took off during the 1950s and most manufacturers switched from using germanium to silicon. Soon they were wired into ever more sophisticated, densely populated circuits that controlled everything from 'transistor' radios to the earliest space satellites. As a child I remember turning the dial of my parents' radio to hear the rhythmic bleeping broadcast by Sputnik 1, the satellite that had won the space race. Even as a nine-year-old, I was in awe of the technological progress that allowed me to connect directly with the first man-made object in space.

But before transistors could really change the world there was an important problem to solve. The robust transistor was too often let down by minute slips of the human hand, as it had to be soldered to other components to make a processor. In the late 1950s, Jack Kilby, working at Texas Instruments, and then, independently, Robert Noyce at Fairchild Semiconductor, converged on similar solutions to the problem, each deciding to build the entire circuit of transistors, resistors, capacitors and connections directly on top of a slice of silicon. Noyce dispensed with wires altogether and deposited lines of conductive material directly onto the semiconductor.[28] Intel's 4004 microprocessor and all of today's chips are direct descendants of these first integrated circuits.

The 4004 chip was not designed to change the world, but it did. At the end of 1969, Intel, the young company where Robert Noyce and Gordon Moore worked, received an order from a small Japanese firm, who wanted twelve different chips for a new range of desktop calculators: separate ones to manage keyboard input, memory, cal-culation, output to the screen and so on. Ted Hoff, the project's chief architect, had a different idea; he wanted to put all the different memory and processing components onto a single, all-purpose com-puting chip. Nothing like this had been attempted before and he had no clear idea how to do it. Work on the new chip stalled, until the clients started to enquire about delivery – and that is when Intel hired Federico Faggin.[29] On his second day in the new job he had to face an irate Japanese representative, who gave him ten months to deliver a radically new type of device. Working under intense pressure and pushing the limits of chip fabrication technology, Faggin built the microprocessor.[30]

The first chips were used in calculators; far more powerful than Babbage's nineteenth-century Difference Engines, they gave many people access to a new and powerful way of working. This was the first step in the democratisation of computing. The Intel leadership had, of course, seen well beyond the calculator – they had built a highly adaptable, self-contained and stable computational device that could be applied to a wide array of technologies.[31] Soon, microprocessors were everywhere: washing machines could offer more flexible and reliable programmes; sophisticated computer games were invented;[32] the control of traffic lights became coordinated and responsive; and trustworthy 'autopilot' controls helped pilots fly their planes more safely. Microprocessors would soon also ignite the robotics industry, which in turn reinvented industrial production processes. One of the key beneficiaries has been the semiconductor industry itself. Over the last fifty years, robotic production techniques have been critical for the sustained improvement of microprocessor capabilities, and the rate of improvement has become so predictable that it is considered to have the properties of a 'law'. We call it 'Moore's Law', since it was first identified in 1965 by Intel co-founder Gordon Moore. Since then, the number of components on transistor-based microchips has doubled roughly every two years.

During my ten years on the board of Intel from 1997, I saw first-hand the explosive growth of the IT sector. In the early days, manufacturers put a great deal of emphasis on designing microchips with properties that they could proudly advertise: greater speed, bigger cache memory or tailored to process media, for example. Then, rather abruptly, the cost and difficulty of adding new features plummeted. For most common purposes, processing power was no longer a constraint, and customers were not prepared to pay more for greater speed. As of 2017, the most powerful microprocessors contained more than 7 billion transistors (nearly 3 million times more than the number in the original Intel 4004). Meanwhile, manufacturing processes have made it possible to shrink the width of a transistor on a chip by one thousand times, to ten nanometres.

When I visited Intel's Californian headquarters in 1997, the offices were grey and formal. Returning in 2017, I found them newly refurbished, with a welcome injection of colour and new artwork on the walls. Inside I met then Chief Executive Officer Brian Krzanich[33] and Corporate Vice President Michael Mayberry. These engineers have both risen through the ranks to these leadership positions. Both are relaxed and friendly, wearing running shoes and jeans. I ask them whether they can continue to improve their microprocessors at such an extraordinary rate. 'We had better be able to!' replies Krzanich, with a laugh. 'Nobody wants to be the CEO at Intel when Moore's Law comes to an end.' Despite frequent reports announcing the imminent demise of the law, Krzanich and Mayberry refute any suggestion that the end is in sight. They are working on designs that will shrink transistors to a width of just five nanometres, or even smaller. Krzanich acknowledges that 'the difference from ten years ago is probably that the time between those [generations of chip] is lengthening ... from about two years to now more like three'. Nevertheless they now achieve more progress in those three years intervals, 'so the slope of the curve stays the same, but the steps are deeper and broader'.

We must never take this most remarkable of engineering feats for granted. As John Hennessy, the Executive Chairman of Alphabet explains, 'What you have to remember is that the amount of resources that have been poured into advancing the semiconductor industry is phenomenal. Very dedicated and focused investment made a series of steps that overcame ... deep, hard engineering and manufacturing problems, re-inventing our manufacturing processes again and again and again.' 'The leaps and bounds haven't been accidental – they have

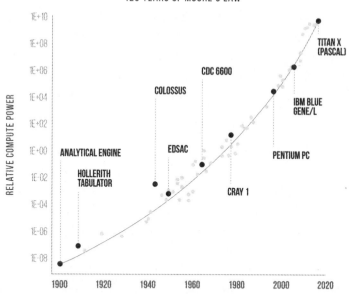

120 YEARS OF MOORE'S LAW

FIGURE 3.4 The power of Moore's Law drives the power of computing, and it's not yet running out of power.

structure to them,' Andy Hopper adds. Part of that structure comes from the International Technology Roadmap for Semiconductors,[34] which has served as a platform where academic researchers, manufacturers, equipment suppliers, government policymakers and other experts can coordinate their efforts in an international and collaborative manner, before embarking on what Hopper calls 'the seriously competitive stage beyond'. Krzanich describes how far his firm has to look ahead in order to continue the delivery of better chip performance. 'From a point where a new material is invented to the time that we use it, if it actually makes that leap, is about twenty years.' All of this has to be done without any reduction in quality and without increasing costs for the consumer. None of this progress is automatic or inevitable. As Hopper emphasises, it is this dramatic and sustained advance in the production of semiconductors that has allowed the technology sector to rise and dominate the global economy. 'The scale of the computer industry, as indicated by market capitalisation, is bigger now than financial services and energy,' he observes. 'The resources available are extraordinary.'

Even if reports of the death of Moore's Law have so far been exaggerated, innovators are looking ahead to new forms of computation that will not be based on silicon transistors. Quantum computers could, some claim, help us compute, and therefore think, in completely new ways.

THE QUANTUM COMPUTING CHALLENGE

At the heart of conventional computation are bits that exist in a binary world, represented only as '1' or '0'. Quantum computers use 'qubits' which rely on particles that, at the quantum scale, have the property of being able to exist in an inherently undetermined state. Electrons, for example, can spin either 'up' or 'down', or in an undetermined combination of these two states,[35] a property called quantum superposition. The cat in Schrödinger's well known thought experiment is trapped in this state – it is simultaneously dead *and* alive, until an observer opens the sealed box that surrounds it, at which point it can only be dead *or* alive. This may defy our common sense-based experience of the world, but engineers can now build qubits with precisely these properties.[36]

Computers constructed from interlinked qubits could be exponentially faster than their conventional counterparts. Conventional computers have to execute each step of a calculation sequentially, but a quantum computer can process multiple calculations in parallel. Quantum computers are not just faster; they also have the potential to solve problems that cannot be practically addressed by conventional computers. A processor of several thousand qubits could, for example, process the complex mathematics needed to defeat the encryption strategies used by today's most secure banking and communications systems – something that today's conventional computers could never do, even if they were left running flat-out for centuries. Manchester University Professor of Computer Engineering Steve Furber points to the vast disruption that one of these machines

FIGURE 3.5 Dead *and* Alive: more than fifty ways to skin Schrodinger's cat, with IBM's 50-qubit quantum computer.

could cause, in the wrong hands. 'That, to me, does not sound like progress,' he says.[37]

There are, though, some more positive applications. Quantum computers could also be used to simulate the states of matter at the atomic scale, a task that far exceeds the capability of conventional computers. Instead of spending huge amounts of time and money experimenting with new materials with new properties, researchers could instead examine millions of 'digital twins', virtual forms of different hypothetical materials – for example, pharmaceutical substances built to target particular diseases. As Intel's Mayberry explains, these capabilities do not yet exist but are the 'kind of "over the rainbow" things that people are pointing at'.

Many companies and research organisations are putting a great deal of effort into developing useable quantum computers. The engineering challenges are substantial: the quantum states of qubits are extremely fragile – any flicker of heat or light destroys their ability to compute, and most must operate in shielded environments, at temperatures a fraction of a degree above absolute zero. Just as Schrödinger's cat resolves into a single state, dead or alive, when someone looks inside its box, qubits also lose the property of superposition when their value is measured directly, which means that all readings must be taken indirectly, adding to the complexity of the system and making checking for errors very difficult. Despite the obstacles, Mayberry is optimistic: 'We think that we can scale it,' he says. Others, however, are not so confident. 'I'll be delighted when it comes along, but I think it is like [nuclear] fusion [power] … always twenty years away,' says Hopper.

There is a popular belief that stable and effective quantum computers, if and when they are available, could be used to solve every computable problem. According to Mayberry, this is unlikely to be true, for two reasons. Firstly, the execution of quantum computation is inherently probabilistic, which means that it returns probabilities rather than definite answers. This can be a powerful feature, but more definitive solutions generated by conventional computers are often needed. 'It's never going to be the thing that sits in a data centre, optimising your search,' says Krzanich. The second reason is more practical. Vast amounts of information must be delivered to and read from quantum processors, a process known in computing as 'input/output' or 'I/O'. 'Even if my computation is fast, I'm limited by the amount of time it takes to put in and get out the information in the first place,' explains Mayberry. All of the auxiliary systems needed to serve the quantum processor will be controlled by conventional computers. For this reason, Krzanich believes that quantum computing is not going to replace Moore's Law or conventional computing, but is rather 'going to be an adjunct to it'.

A unifying aim for the engineers of quantum computers is the pursuit of 'quantum supremacy', the still-elusive point when a quantum computer solves, in a practically useful time frame, a mathematical problem that even the most powerful conventional supercomputers could not solve. For now, useful quantum computers only exist in the imaginations of engineers and physicists. 'The physics is really interesting. Whether or not this will scale to a sufficiently large and general-purpose technology remains open,' concludes Hennessy. We still live in a world where almost all computations are performed by silicon-based microprocessors that execute operations in a binary fashion.

INCREASINGLY PERSONAL COMPUTERS

Steve Furber was a principal designer of the Acorn RISC Machine, the first iteration of the chip architecture that now controls tens of billions of smartphones and Internet-connected devices around the world.[38] When asked how the engineering of the first microprocessors changed the course of computing, he says, 'within ten years it had enabled the personal computer. It took the computer out of the machine room where it was only ever touched by men in white coats and put it into the hands of everybody.' This was a profound shift, and Furber thinks the key factor was cost: 'As with so with so many engineering technologies, if you can transform the cost, you can transform the scale of use.' In 2017, more than half the world's population accessed the Internet via some form of personal computer.[39]

By the early 1990s, companies such as IBM, Apple, Atari and Sinclair were deploying hundreds of thousands of microprocessors in home computers each year. Now they are everywhere – in our smartphones, in watches on our wrists, in medical devices, cars, toys, household appliances and billions of other sensors. They are always on, always connected and always computing. When they are connected and respond to each other, they become what is known as the Internet of Things. Could it be possible that our brains could simply become a part of this wider network? I will explore this futuristic possibility in Chapters 4 and 10.

While microprocessors democratised access to computing, they also instigated the next big transition in the computing sector. Prior to this point, hardware engineers had always taken the lead – but now software engineers had to up their game. If ordinary citizens were actually going to buy these machines and use them to work and think in new ways, software would have to get very much better. In 1970 Edgar Codd, a researcher at IBM, wrote that 'future users of large data banks must be protected from having to know how the data is organised in the machine'.[40] He then set out his vision for how computer databases should be built and used. Far from being a dry and irrelevant academic theory about data management, Codd's insights would soon change the world.

Huge sectors of the modern economy are built around collecting, combining and analysing data. Think of Google and Facebook – data is their raw material, their capital and their product. Databases may seem impossibly dull, but it is no exaggeration to say that the rise of the digital economy, and much of the growth of more traditional economies during the same period, was only possible because of the foundations laid by Codd. When he wrote his seminal paper, computers were not yet in people's homes, though most large corporations already relied on them, and managing the data they stored on these mainframe computers was becoming something of a headache.

Codd's solution, which he christened 'the relational database model', eliminated the idiosyncrasies of existing systems. He was the first person to devise a logical and overtly engineered approach to data management.[41] Despite the clear advantages of this new database model, his bosses at IBM rejected the idea of developing it as a commercial product,[42] but while they dithered, a young Silicon Valley-based entrepreneur called Larry Ellison, the founder of Oracle, spotted his chance to build a software package to create and manage relational databases. In 1979, Oracle Release 2 (there hadn't been a Release 1, because the developers feared that the name would make the product seem immature) was launched. It proved popular with customers and rapidly swept through the corporate world, giving businesses an easy way to store, analyse and retrieve all kinds of data, from inventories to mailing lists. In 2017 Oracle was the second highest-grossing software company in the world, after Microsoft.

The eventual success of the relational database model underlines an important engineering principle. Codd had sought to impose clear standards on electronic data management – and while they may appear boring and conservative, they can be relied upon without being questioned and they can also make creativity possible within an engineering framework. For example, the simple invention of standardised screws by Henry Maudsley and his screw-cutting lathe in around 1800 almost certainly accelerated the Industrial Revolution – imagine engineers attempting to build a huge, multi-component system like a space rocket without standardised, quality-assured screw threads.

The technology entrepreneur Suranga Chandratillake sums up the influence of the relational database on the world of commerce: 'One thousand tasks that we used to do manually, in paperwork, logbooks and reports were rapidly disrupted and turned into computing processes.' The key consequence of this automation was to allow businesses to grow and operate at scale. Finance, logistics and markets could be managed across the globe. When coupled with the Internet, databases gave consumers instant access to new banking services, online shopping and many other web-based services that we now take for granted. Chandratillake concludes: 'Theoretically, you could have done all those things before, but it would have relied upon millions of pages of paper and people shuffling through them constantly. [The relational database] for me is pretty iconic among that first wave of software solutions.'

Even as computers were providing new services and business opportunities, many feared that they would also create significant unemployment. However, just as the replacement of physical tasks by machines caused unjustified concern, the same pattern has emerged here – computers have automated part of our mental work, without replacing us entirely. For example, computers have been used for at least a decade in the legal industry's 'discovery' process, digging through company records to find documents that are relevant to specific legal cases. This replaces expensive and time-consuming work that was previously done by lawyers and paralegals and, in fact, the software often does a better job. However, contrary to expectations, it has not led to widespread unemployment in the legal profession; instead, a profitable new branch of the software industry has developed, and the number of lawyers and related workers employed in the US has grown at a faster rate than that of the labour force as a whole.[43] The same trend has played out in various other sectors, shifting labour to activities where it can add more value – a phenomenon that we explored in the last chapter. Computers have helped the world grow enormously, by creating both work and opportunities.

The relational database was an invention that removed a huge amount of mindless record-keeping, rote learning and arithmetic from many jobs. In today's world, corporations and governments rely on vast quantities of data, which exists in many forms and is in constant flux. As a result, the highly structured world of the relational database has, in many cases, become too restrictive.

The computer scientist Onnig Minasian explains this problem with an analogy: 'Consider the automobile as a complex object. When you use an automobile, you use the entire thing. And when you store the automobile you don't take the wheels off, the steering wheel off, the engine apart. You store it as an object. And the limitations of the relational model are that it just doesn't lend itself very well to handling complex [computational] objects. Relational databases have to disassemble and reassemble everything.' One recent solution to this problem is the non-relational database. These systems allow for distributed, dynamic and flexible data storage and can be used to absorb the digital details of our economies and daily lives. These are the databases that companies, governments and intelligence agencies use.

In the last decade, a new form of distributed database has been invented: an electronic ledger, which records transactions in a tamper-proof and decentralised way. It is called the blockchain, and its most ardent advocates believe it could change life even more dramatically than the more traditional incarnations of the database did.

THE PROMISE OF BLOCKCHAIN

In 2008 an enigmatic programmer (or group of programmers), working under the pseudonym of Satoshi Nakamoto, published a technical paper on an obscure online cryptography bulletin board.[44] In it, he described the workings of the blockchain, and his plans for its first implementation: a new digital currency called Bitcoin. The system would automate the process of recording and verifying financial transactions, without the involvement of a bank or any other intermediary. Transactions would only be approved when every computer in a network of hundreds or thousands of machines received formal proof of their authenticity. Discrete blocks of transactions would then be saved on every computer in the network and cryptographically 'sealed'. If anyone attempted to manipulate the record of transactions, the cryptographic key would no longer work, and because the encryption applied to each block depended on the previous block, they would be linked in a *chain*. Any attempt to change, copy or delete past records would disrupt the entire blockchain, immediately exposing any interference. Thanks to these properties, blockchains would be safe, distributed ledgers.

In January 2009, Nakamoto laid the first block in the first ever blockchain. He improved his system over the two years that followed but then, just as his project was gathering momentum, he disappeared. Since then there has been a great deal of speculation about who this mathematical genius is or was, and why he kept his identity such a secret. The blockchain system engineered by Nakamoto represents a completely new way to record and authenticate a whole range of human interactions – crucially, it allows people who do not know each other and have no reason to trust each other to create a robust and mutually satisfying record of ownership and obligation. For the first time, people can do this without relying on the trustworthiness of a bank, government or other arbitrator. Cryptocurrencies may well turn out not to be the most enduring and important application of the blockchain – they could potentially change how property rights are transferred, for example, with 'smart contracts', which would pay invoices only on when a product or service had been delivered. Medical records could also be stored in a distributed and yet tamper-proof way.

At a recent meeting in Washington DC, impact investor Andy Karsner proclaimed that 'Blockchain is as significant as the Internet.' However, Vint Cerf, co-inventor of the Internet, is not so convinced. 'Blockchain is an interesting technology,' he says, 'but let's be careful not to attribute to it the solution to everything.' Cerf thinks the mechanism for reaching consensus is too slow and too energy-hungry, and that the system lacks mechanisms to ensure that the contents of databases are secure.[45] Andy Hopper echoes this view; he reminds me that 'chains of blocks' have been 'a fundamental underpinning of computing since the word go'. 'If you just change the words around, all of a sudden it becomes less effervescent and perhaps seen for what it is – a nice innovative step. It will get us some good stuff, but it is not the solution to all the world's problems.' Just as no one yet knows the identity of Nakamoto, no one yet knows what impact blockchain will have our way of life.

COMPUTING IN THE CLOUD

The rise of distributed database systems has proceeded in step with the rise in access to distributed computing resources. This has been a gradual process. First, data was moved from storage on mainframe computers to remote servers. Then the Internet opened up the possibility of truly distributed storage. Now, with the spread of high-speed connections, computers that are connected to the Internet are not just useful places to store data – they also give us access to immense, so-called 'cloud-based' computing power.

The transition to cloud-based computing has demanded a psychological shift in how we think about our computing operations and data storage. During my time as CEO of BP, for example, I recall the suspicion that greeted my decision to use the Internet for corporate activity and to consolidate the company's computing power and databases into remotely located centres. The concerns echoed early twentieth-century worries about sharing telephone lines; many thought that their privacy would be violated by others listening in to their conversations. Similar challenges remain but I suspect that, over time, unease about cloud computing will seem just as outmoded.

I visited one of Google's newest office complexes in Sunnyvale, California, to learn more about the future of distributed computing. Ducking in and out of the bright Californian sunlight, I met Fei-Fei Li, Stanford Professor and Chief Scientist at Google Cloud, who calls cloud technology 'the biggest computing innovation humanity has done'. Soon, she thinks, computing power will never be a limiting factor again. 'It literally is an endless number of computers ... you don't even need to know where they are,' Li continues. She describes how the cloud 'reaches every single industry: from energy, to commerce, to financial services, to health care, to agriculture, to manufacturing'. However, before cloud computing can fulfil its huge potential, engineers must work out the details of how physical computer servers, software platforms and distributed databases and software tools can integrate seamlessly. Another real and growing challenge for cloud computing is cyber security, an important topic that we will return to in Chapter 8. As ever more software is virtualised and run on remote servers, ensuring that everything is secure against meddling by malign actors or theft becomes very much more difficult. When engineers answer these challenges, computing power will become just another invisible aspect of the hidden infrastructure upon which the world relies.

THE DEEP ROOTS OF DEEP LEARNING

Access to powerful computers and huge, flexible stores of data are driving another great wave of computing innovation that is already starting to change what we can do with computer programs. The dominant model of software engineering since the middle of the twentieth century has been to construct algorithms according to inflexible rules. This means that the way in which an algorithm performs is predictable in almost any set of circumstances. The pioneering Ada Lovelace had this idea in mind back in 1843, when she wrote, 'The Analytical Engine has no pretensions whatever to originate anything

… It can follow analysis; but it has no power of anticipating any analytical relations or truths.'[46]

Since the beginning of the digital age, people have dreamed of computer systems that can think and reason in more flexible ways. What could be done with machines that can improvise and react to new circumstances, and might there ever be machines that can look out into the future and predict what will happen next? At first, attempts to build so-called 'artificial intelligences' tried to define the rules and logic that underlie intelligent human behaviour, a method that is sometimes referred to as the 'symbolic reasoning' approach to artificial intelligence, or 'AI'. To make programs that could translate languages, for example, computer scientists asked linguists to provide the rules that define a language, which they could then reproduce in code. These were so-called 'expert systems' and they were able to perform specific operations well. NASA, for example, relied on such systems to make flight decisions during the space shuttle missions in the late 1980s and 1990s.[47] Almost all the systems that control industrial robots and autonomous vehicles are still based on symbolic reasoning, but this way of doing things has proved to be too limiting for broad application. For example, languages are too complex and nuanced to be reduced to a tidy set of rules; almost every linguistic convention has many exceptions.

In 1950, Alan Turing sketched out a future for artificial intelligence, proposing that the best way to build a computer that could think like a human would be to engineer it in an immature form. 'Instead of trying to produce a programme to simulate the adult mind,' he wrote, 'why not rather try to produce one which simulates the child's?'[48] We gather knowledge and insight gradually, from others, from reading and from our experience of the world. With all this learning we build up a model of how the world works, which then directs our thinking and behaviour. Turing's vision was clear, but implementing it has proved to be very difficult. That is now starting to change. 'After fifty-plus years of investment and forty-five years of failed promises to really capture this, all of a sudden, "boom", it is happening,' says John Hennessy, his voice rising as he emphasises his point. 'There are some times in engineering where you see a real turning point,' he continues, 'where something really essential is discovered, like the first transistor or the first integrated circuit.' He believes that we are now witnessing the next pivotal moment in the course of computing history.

In around 2017 a computer algorithm called AlphaGo beat all the world's best human players at the ancient and complex board game Go. The most surprising thing about its victories was the way in which the algorithm won. It was given the rules of the game, but it was not given any winning strategies.[49] Instead it 'learned' by accessing a huge database of Go games, and then playing against itself hundreds of thousands of times, exploring all possible successful strategies. As it was successful it adjusted its own parameters, of which there are many millions, so that it steadily improved its ability to win the game.[50] This approach, reinforcement learning, comes closer to Turing's notion of a machine that learns than anything that has been built before.

The power of machine learning has been vastly improved by the use of neural network algorithms that loosely mimic the function of the brain's neurons,[51] which are arranged in a hierarchy of layers. Layers that are higher up in a neural network algorithm's hierarchy deal with increasingly sophisticated concepts or aspects of perception. 'Neurons' in each layer are designed to respond to specific features of the data presented to them and only react when they receive the stimulus relevant to their layer. For example, if a neural network is designed to recognise the content of photographic images, the lowest layer of neurons may respond to simple features such as edges, colours or basic shapes. The higher layers respond only to increasingly complicated shapes and patterns. The final layer, at the 'top' of the network, will respond only when it recognises a specific image, such as a particular face or type of animal. The term 'deep learning' simply refers to the use of neural networks with a large number of intermediate layers. Fed with enough data and feedback on whether it is reaching an objective, a deep learning algorithm gets progressively better at its task.[52] With enough training, these algorithms have the potential to identify the real meaning of written words, spoken languages, medical data and much more besides.

It is popularly believed that machine learning is an entirely new and transformative approach to computation, but that is not actually the case. The mathematics upon which it depends is not new, deriving from the discipline of numerical analysis, the study of algorithms that use numerical approximation to solve analytical problems. The foundations of all these techniques were laid in the seventeenth and eighteenth centuries, by such luminaries as Isaac Newton, Joseph-Louis Lagrange and Leonhard Euler. During the nineteenth century, Charles Babbage, Carl Friedrich Gauss, Augustin-Louis Cauchy and Karl Pearson, among others, added new insights. Then, during the Second World War, the field of operations research matured and was used to find optimal solutions to complex problems that involved many variables. This meant, for example, finding the most efficient routes and formations for supply convoys, streamlining procedures in crucial manufacturing plants and allocating finite military resources to multiple battlefronts. To give an example of the kind of gains that were achieved, the average number of anti-aircraft artillery rounds required to bring down an enemy plane fell from more than 20,000 at the start of the Battle of Britain to fewer than 4,000 in 1942. It was also during this war that Walter Pitts and Warren McCulloch, two University of Chicago mathematicians, wrote the first mathematical description of a neural network, the predecessor of all today's deep learning algorithms.[53] The 'backpropagation' algorithm that is today considered key to the utility of artificial neural networks was first conceived in 1960. Zoubin Ghahramani, Cambridge Professor of Information Engineering, points out that 'There's rarely anything new under the sun ... Fields like AI and machine learning are constantly building on other fields: sometimes reinventing things; sometimes extending; sometimes looking at old ideas with a new perspective.'

There is no magic to neural networks or deep learning, ideas that are no more complicated than the sophisticated application of statistics. The key factor that unshackled the potential of neural networks during the course of the last decade was not an algorithmic breakthrough – it was the increased availability of big datasets for algorithms to learn from, and increased access to the gigantic computing power needed to process such large amounts of data. And, as AI expert Stuart Russell cautions, better computers alone cannot solve all our problems: 'If you make the computer faster, you just get the wrong answer more quickly.'

Services based on these machine learning approaches are becoming ever more common and sophisticated. When your smartphone is activated by your face or responds to your voice, it is using machine learning. Recommendations by online retailers, such as Amazon, of what you might buy are based on learning from your past purchases. And John Hennessy invited me to think about a dermatologist, whose job it is to diagnose skin lesions. 'He's been trained to look for skin cancer, but how many lesions does he look at in a lifetime – ten or a hundred thousand, perhaps? ... Now suppose you get hold of 10 million images from around the world, which have been classified. This one was cancer; that one wasn't. I can build a system that will take in all that knowledge,' he says. 'You cannot get that knowledge into the head of a doctor in a lifetime.' But now, with the help of a machine learning algorithm, the dermatologist can integrate all of that information and make a better diagnosis. I experience this myself every few months when I go to my ophthalmologist about my glaucoma. After having my retina scanned by a technique called Optical Coherence Tomography (OCT), an algorithm compares the data from my eyes to that from thousands of other scans, which have already been assessed by expert physicians. The algorithm is trained to spot subtle patterns that reveal important aspects of eye health. The computer's judgement helps my ophthalmologist decide on the best course of treatment, but it does not replace him.

This application of machine learning is poised to transform eye care. There are currently many more OCT scanners than there are trained ophthalmologists who can interpret the scans; the high street opticians who perform most of the scans have no option but to refer everyone with an apparent anomaly to specialist eye clinics, despite the fact that many of these cases turn out to be false positives. As a result, eye clinics are overwhelmed and people with genuine problems often have to wait weeks to receive care. And, as ophthalmologist

Pearse Keane says, 'If we intervene earlier, we have much better outcomes. If it was a family member of mine, I would want them to receive treatment within forty-eight hours.'[54] If an algorithm could offer a rapid and reliable pre-diagnosis, it would remove the burden from the system and stop treatable eye problems from developing into serious visual impairments.

IDIOT SAVANTS AND BLACK BOXES

'Neural nets are an absolutely phenomenal technology,' begins Mark James, Chief Technology Officer of Beyond Limits, a cognitive AI company spun out of NASA's Jet Propulsion Laboratory.[55] 'But at this point,' he continues, 'the AI field is really plagued by the concept that artificial intelligence and neural nets are almost indistinguishable.' He explains how AI researchers during the 1970s made boastful claims and 'got a black eye for their arrogance' – he believes that this set back progress across the whole field and is wary that expectations are once again being over-inflated. Powerful as neural networks are, James points out their limitations. Firstly, they cannot help in truly novel situations. 'If it hasn't been seen before, and seen in many different forms, they simply don't solve the problem.' Secondly, training neural networks is a delicate art. 'If you overtrain them, they become too specific to a particular dataset and then they miss variants of it; but if you undertrain them, you get errors in the output.' Thirdly, neural networks are very computationally intensive; this is being helped by new chips with specific functions, including 'Tensor Processing Units' (TPUs), 'Graphical Processing Units' (GPUs) and 'Intelligence Processing Units' (IPUs). These provide the high degree of parallelisation that is required for the rapid tensor multiplication operations that neural networks depend upon but the process is still far from efficient, 'which is why Google has warehouses full of computers to provide that function', explains James. Finally, and perhaps most troubling of all, they are, in James's view, 'essentially idiot savants'. This means that they may 'do a phenomenal job of making observations, but you really don't know why and you cannot invert them'. In this respect, machine learning algorithms are very different from the programs that I created to analyse oilfields in the 1960s and 1970s. Machine learning programs are sometimes described as 'black boxes', with all the opacity that the term implies.

Professor Hopper is uneasy about black boxes, particularly if they appear to reach conclusions that directly affect people's lives. His overriding concern is that 'if [an algorithm] is a black box, you don't really know how it works, so you can't apply policies to it. And therefore you might inadvertently let it do the wrong things.' For example, if in the future there is a machine learning algorithm that diagnoses cancer and makes treatment recommendations without the need for expert human oversight, how comfortable should a patient feel about heeding its advice, if neither the algorithm nor the doctor can explain the rationale behind the advice? As algorithms begin to make more consequential decisions about insurance premiums, credit-worthiness, employability and how best to drive an automobile, the same question will arise. How valid are judgements that cannot be explained in terms that people understand?

Hopper points to another risk. He imagines that an artificial intelligence system is used to make parole decisions for prisoners. 'What happens if it decides that all white people get to go free and all black people have to stay? It didn't mean it like that, but that's what happened.'[56] This is one example of algorithmic bias; the subtle decisions coders make can influence the performance of a system that is based on machine learning. Algorithms can also reinforce biases when they learn from historical datasets, such as past sentencing decisions, which are based on historic prejudices.

Google's Li is worried that her mission to democratise AI and machine learning could be derailed by what she describes as a 'lack of diversity in technology'. 'I think this issue is much more important than "Terminators", yet no one talks about it,' she tells me. 'We are talking about a technology that carries so much human value,' she continues 'and we don't even have the right representatives in this technology.' Indeed, employment at high-tech firms in Silicon Valley is disproportionally skewed towards white males, while women, African Americans and Hispanics are under-represented.[57] This matters because, as Li makes clear, 'It's not enough to just have that technological dream, it's much more important to know how to harness this technology, how to use it in a benevolent way.'

Li's concern crystallises the importance of inclusive leadership. Not only do more diverse teams to a better job, but by drawing from a deeper and wider pool of experience they also reduce the risk of embedding engineering bias into our most important technologies. As we ask algorithms to make ever more decisions that affect our lives, they must embody a broad and inclusive set of human values and reach conclusions that we feel are naturally fair. John Hennessy agrees: 'If we as a human population are ignorant enough to leave all kinds of decisions about who lives and who dies to computers when we know that those computers were programmed by people who were inevitably fallible, then we get what we deserve.' As he makes clear, engineering leaders, funders and regulators must work together and take decisive action to ensure that these systems incorporate a wider range of moral values than they do today.

All of this leads Andy Hopper to say, 'there is a principle that needs to be seeded here: in a professional engineering framework, you must not have black boxes'. Steve Furber is also concerned about the black box nature of a neural network algorithm. 'You can't analyse [such] a system atom by atom. You can only establish its capabilities by testing.' But, as Furber explains, 'that's also true of quite a lot of engineering, but in computing circles we're used to being able to prove things about the capabilities of software mathematically. In the case of neural networks that's just intractable.' Ghahramani believes that 'for some applications, we're perfectly happy with black boxes that just work reliably'. When it comes to systems that target advertising or suggest movies to watch, a mistake caused by an opaque system is generally inconsequential. However, for more important decisions, Ghahramani is exploring ways to make algorithms more comprehensible[58] and 'to build parallel systems whose job it is to explain what the black box did. In theory you can even train those systems to give a good explanation.'

Mark James thinks that part of the solution to the 'black box problem' is to fuse the more opaque neural network-based machine learning with symbolic reasoning approaches – the latter, he says, will be able to 'operate like an orchestra leader', collecting the results from the numerical algorithms together, to form higher order representations that will make sense to people. With this aim, he is building AI algorithms that are 'both trained and educated', rather than simply flooded with training data. Tong Zhang, head of the AI lab at Tencent, one of China's leading tech firms, agrees that more interpretable AI systems are needed and that more sophisticated 'modularised' approaches like James's may be the best way to achieve that aim. He warns, however, that most interpretable AI systems have so far 'lacked the performance of the black box'. Ghahramani adds a further note of caution: 'The explanation [that these systems offer] might seem plausible, but might not be the real reason. When a decision comes out of a very complicated system that's been trained on massive amounts of data, it's not even clear what an explanation is.'

It seems to me that Ada Lovelace was wrong when she claimed that computers would never 'originate anything'. Artificial intelligence may not yet be able to write books, but it can come up with novel and surprising results and forecasts of the future. For example, people who play Go against algorithms are often surprised by the moves that they make. Even so, Li points out that such achievements are still 'very, very niche'. 'Games like Go are virtual worlds with completely perfect, clear rules. They are finite worlds.' Artificial intelligence is still an emerging field, and Li reminds us that 'it is going through a huge hype wave, and people extrapolate overly zealously'. Tim Westergren, founder of the Pandora online radio station, points out that computers are still clumsy in their attempts to classify music, let alone compose it – as a result, the station employs human musicologists to curate its musical content. Hennessy confirms this missing 'skill'. He says, 'Look at computer music. It's okay, but it's not Beethoven. Computer art is okay, but it's not Picasso or Matisse. It is missing something … I think that a little bit of humility is probably good here.'

Nothing computers have done so far can compete with our brains in terms of originality or flexibility. They are not poised to outwit us, and perhaps striving to make them think more like us is the wrong approach. Instead, machines will always be most useful to us precisely *because* they think in very powerful but distinctly unhumanlike ways. Babbage's numerical engines executed arithmetic with inhuman accuracy. EDSAC could process datasets that far exceeded the capacity of the scientists who used the computer. Codd's relational databases allowed us to store information in perpetuity and to access it immediately. Blockchains provide a way to instantly verify authenticity with thousands of partners. Neural networks can learn to identify patterns that are invisible to us. However, machine learning still has a long way to go – for an algorithm to recognise a cat, it has to scrutinise millions of images of cats, while a toddler, shown just one cat, will recognise cats for ever. Computer-assisted thinking complements human intelligence, but it does not make it obsolete. And when it comes to creative thought, I believe that a device that is very much simpler than a computer is still our most effective aid.

THE PEN

The pen is a symbol for what is unique about our brains. For as long as we have been humans, we have been making marks. At first we drew on the walls of caves. Ancient Egyptians and Mesopotamians started to write their words down and, two thousand years later, court officials in China started using paper. Over the centuries, brushes, quills and reeds gave way to pencils and fountain pens. My small collection of Japanese lacquer pens is among my favourite possessions. Each is a miniature work of art, made by an artist who has spent decades mastering the craft. They are decorated exquisitely with intricate patterns and scenes of nature, and when I use them, they turn the mundane act of putting pen to paper into an inspiring ritual. Taking the time to select the pen and colour of ink that I want to use that day and making the first marks on a pristine sheet of paper have become integral to my thinking process.

For me, the act of writing with a fountain pen is an empowering reaction to an age in which computational devices and software services are everywhere, promising ever more speed, convenience and instant gratification. Handwriting forces us to paraphrase and synthesise rather than to simply parrot, which helps knowledge to stick. It is not just me who believes this – cognitive scientists recently showed that students retain information better if they take notes longhand rather than by tapping at a keyboard.[59] Moreover, writing is an intimate and engrossing act. With a pen in hand, I must commit to the words that I put down on the page, and the momentary resistance between the idea and its recording gives me space to think. It is the only reliable way I have found to plan what I want to say and to say it with economy; the first outlines of this book were written on plain white paper with pen and ink.

Across history, the vast majority of our most humane and moving works of art, literature, science and politics have been sketched, painted and written by hand. And from Babbage's mechanical

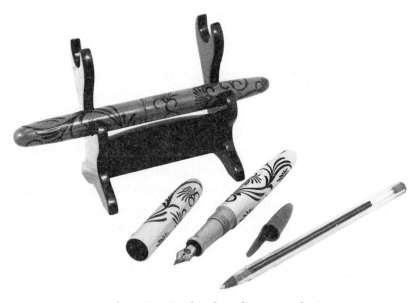

FIGURE 3.6 Namiki vs Bic. Handmade vs factory-made. Precious vs ubiquitous. However different, both are aids to capturing thought.

drawings to Federico Faggin's designs for the first microprocessor, many of engineering's most powerful inventions started as freehand diagrams; keyboards are convenient, but they limit the diversity of symbols and marks that we can use to express ourselves. New 'user interfaces', including voice and gesture recognition, are improving quickly, but for the time being even a finger on a highly responsive touchscreen cannot match the simultaneous precision and freedom of putting pen to paper.[60]

Ordinary as the pen may seem, it was only relatively recently that it spread around the globe and across the socio-economic spectrum. The Museum of Modern Art in New York describes the elegant Bic Cristal ballpoint pen as a 'humble masterpiece'. At the heart of its mechanism is a heat-vitrified tungsten carbide ball bearing, precision milled to ensure a smooth writing experience. A tiny hole in the barrel equalises pressure, which means the ink flows at a constant rate. The transparent polystyrene barrel is hexagonal, making the pen both stronger and easier to grip. In its essence, the Cristal is a triumph of the engineer's ability to produce a product that is reliable and retails at a minuscule price. The Cristal's inventor Marcel Bich first encountered a ballpoint pen in Argentina, where the Hungarian Jewish inventor Ladislao José Bíró had taken refuge during the Second World War. Bich bought the rights to Bíró's patented technology and returned to France to develop the product that he would eventually launch in 1950.[61] In 2005, Bic sold their one hundred-billionth pen.

Automation has always changed the nature of the jobs done by humans. Current computing advances are no exception and while they will continue to change workplaces in many ways, they will not make human decision makers redundant or lead to mass unemployment. They will, however, provide many ways to augment our thought processes and to compensate for our brain's inherent limitations. We already use them to outsource our memories, radically speed up calculations and broadcast our views around the globe. Computers give us free access to ideas and enable us to build on them. After decades of development, the combination of ultra-fast computing devices, cloud storage and new algorithms have introduced long-heralded practical applications of machine learning and artificial intelligence. Though these machines appear to be learning, they do so in ways that embody the biases of their designers and the peculiarities of the experiences that they have learned from. As things stand, they behave as 'black boxes', but their processes must be open to scrutiny, or we will lose control of them. If we cannot follow their logic or spot the flaws in their decision-making, we cannot know when they have reached the wrong judgement, and if that happens we will have little hope of pre-empting unintended consequences. When it comes to making consequential decisions on the health or well-being of people, the very last step in the process must be the scrutiny and judgement of a human. That last step must never be delegated to a machine.

I am in no doubt that computers are the most powerful tools that we have built. They support us in every domain of life, from the creation of works of art, to the education of our children and the recycling of our waste. Their transformative potential is obvious, but I am equally clear that they remain our servants, and not our masters. Zoubin Ghahramani expresses this most clearly. Discussing artificial intelligence algorithms, he says, 'From my point of view, these are just tools, ultimately in human hands for doing things that are useful for people.' We have the chance to use them to elevate our thinking and thereby continue, or even accelerate, the progress we are making towards a better world.

4

Connect

Edward Snowden made global headlines in 2013 when he revealed the extent of the secretive web of digital surveillance in which the US government wrapped its citizens. He subsequently appeared in the play *Privacy* at New York's Public Theater. The protagonist, played in that production by Daniel Radcliffe, is on a quest to understand himself, his motivations and his ability to form human relationships, in a world in which networked technology increasingly defines and records the extent of his interactions. The play follows him as he struggles to make sense of this new way of living; Snowden appears on a live video link and reinforces the point that, as inhabitants of a world that is totally reliant upon the Internet, we have already conceded any notion of genuine privacy. Despite the comfort of my theatre seat, I was struck by twinges of unease as I watched; I knew rationally that much of what Snowden and his fellow actors were saying was true, but emotionally I still wanted to reject it. Privacy has not yet been eliminated, but it is under serious assault – we may end up in a world without privacy where, with some important exceptions, people are vulnerable to manipulation and no one is truly free.

The Internet undoubtedly makes my world more convenient. I can instantaneously contact friends, family and colleagues, wherever they are on the planet. Shopping, meals, taxis and entertainment all come to me. I can tap into a huge portion of the knowledge, art and insight that humankind has accumulated over the ages, at no obvious monetary cost. Crucially, the Internet also lets me predict the future in a thousand small ways; it shows me the most efficient route across a crowded city, and it knows what I would enjoy reading or watching next. It even tries to complete my thoughts for me and to pre-empt my online search queries.

But back on stage, the play examines the bigger story about how the Internet is changing the fundamental aspects of what it means to be human. As we watch, we are being watched. As we learn, others learn about us. Our patterns of behaviour, our movements around a city and our use of digital devices are all deeply revealing. Corporations and governments hoover up the trails of data that we inadvertently, and even casually, leave behind, using these insights to fill vast 'data lakes' from which they try to deduce our next move, our next click and our next desire.[1] A new industry has emerged that mines our private worlds to gain financial or political influence. We may gain some measure of convenience or security, but ultimately, we are robbed of our privacy. In a world where our actions, both online and offline, are increasingly seen by state and corporate actors, how free are we to speak our minds or voice dissent? When I was a child, my mother, as a result of her experiences in Auschwitz, always counselled me, 'Don't trust people with your secrets because they will use them against you.' That was good advice then, but it now it feels as if governments and companies know our secrets without us revealing them. They are, of course, not reading our minds; however, the perception that they might be doing so gives them great power.

At the headquarters of the US Defense Advanced Research Projects Agency (DARPA) in Washington, DC I watch as a man, who is paralysed from the neck down, manipulates a robotic arm with unerring dexterity. Just by thinking, he can flex or rotate every finger and joint of his bionic hand. A pinhead-sized sensor carrying ninety-six microscopic electrodes sits on the surface of his motor cortex, the part of the brain that coordinates voluntary movements. Not only can he move his prosthesis, he can also feel what he is touching – tactile sensors on his robotic hand produce signals that are delivered, by another implant, to the parts of his brain that process sense information. When he squeezes his girlfriend's hand he can, for the first time since the accident that paralysed him, actually feel what he is doing – it is hard not to be moved by the thought. 'Look how naturally he uses it! It was almost instantaneous. There was no training,' says Justin Sanchez, the Director of DARPA's Biological Technologies Office. He makes it clear that the brain-guided prosthetics he showed me on a video are just the beginning. 'We don't want the incremental result. We want something that's going to change the game,' he says. His next challenge is to mend, and perhaps even augment, the brain's memory capabilities. DARPA's investigators have located sites in the brain where they can implant electrodes before stimulating neurons with electrical currents, which result in measurable improvements

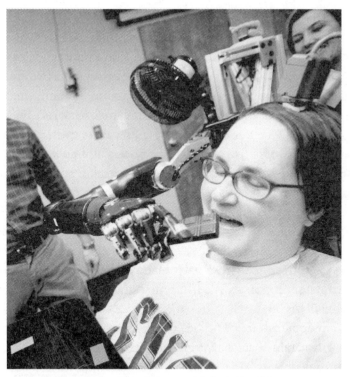

FIGURE 4.1 Thought for chocolate: quadriplegic Jan Scheuermann controls a mind-controlled robotic arm.

in memory function – '20, 30, 40, even 50 per cent improvement, depending on the person', Sanchez reports.[2]

Many predict that direct and seamless connection between our brains and machines is fast approaching as the next frontier in communications technology. It remains too early to know how far these devices will progress – it might be that our most complex thoughts and emotions are impossible to decode. But even contemplating this future reminds us that the content of our communication is much more important than the specific qualities of the devices that link us. We could, for example, use neural interfaces to brainwash people or to control robots at a distance in warfare, but we could also use them to heal mental illness or to become more empathetic by sensing what others are feeling.

THE POWER OF THE NETWORK

As a species we are built to connect – we instinctively crave meaningful social contact and look to share information with others. Language is our original means of connection and it has been crucial to all that we have achieved. In enabling us to communicate, language makes humanity more than just the sum of its parts – it gives us the power to divide our efforts, learn from others and organise ourselves into networks that are far more powerful than any individual person or organisation. The smartphone is itself a powerful tool for communication, which also represents how much we can achieve by connecting with others. Rather than having a single inventor, its design built upon years of innovation and connected the work of thousands of engineers. Connectivity also permits the typical smartphone to be assembled from parts made to detailed specifications in many different countries. And, of course, the device itself exemplifies connectivity.

In an extraordinary end to a meal in California in the year 2000, I watched as Nobel Prize-winning economist Mike Spence demonstrated the power of networks. Drawing directly on the restaurant's paper tablecloth, he reconstructed Metcalfe's Law from scratch. According to this rule of thumb, first expressed by Robert Metcalfe, co-inventor of Ethernet, the value of a telecommunications network grows in proportion to the square of the number of users connected to the network. So if a network doubles in size, the number of possible connections increases by nearly four-fold. More recently, another Internet pioneer called David Reed argued that Metcalfe had been too conservative in his estimation; Reed's Law states that the impact of a social network increases at an even greater exponential rate.[3]

Scaling laws of this kind apply to all information networks, whether or not they are mediated by technology. This explains why human knowledge is cumulative: as more people contribute to a network, more useful insights emerge and add to those that already exist. Ever since our ancestors began to mimic one another before learning how to converse, intelligence has been a collective rather than a private property, which grows most quickly when connectivity is easy. This is why connectivity, leading to communication, has been necessary for all progress. As we will see, every great step forward in social organisation and technological sophistication throughout history has been accompanied by innovations in communication. With agriculture came the written word, first as a crucial record-keeping device, and later as a means of free expression. The ideas behind the European Scientific Revolution and the Enlightenment were only able to take root and grow because of the texts and images reproduced by the printing press. The Industrial Revolution was hastened by the ability to encode words electrically, first in telegraphs and then with telephones, radios and, eventually, computers.

By first building computers and then linking them together to form the Internet, we have radically increased our powers of communication. And, as I saw at DARPA, we are improving direct communication between the brain and machines – computers are beginning to understand our words, interpret our gestures and read the images that we make. Machines also share information with one another in sophisticated and often unsupervised ways. As more devices start to actively participate in our information systems, the capabilities of those networks will grow.

The billions of people using smartphones, personal computers and tablets can be linked together in an apparently infinite number of ways. Many of these links result in traffic that is worthless noise, but if only a tiny fraction of these interactions is useful or productive, the opportunities created are significant. The Internet offers new ways to trade and exchange goods and services. Scientists and engineers can make more rapid progress by collaborating and pooling their knowledge. Educators can reach ever more students. Communities can form, regardless of geographical constraints, to provide support for minority groups and campaign for civil rights.

But there is also a dark side to this new world. Many people spend many hours interacting virtually with only a few others and, as a result, society is becoming more fragmented. Access to unlimited information tricks us into believing that we understand the world better than we actually do. Perhaps smartphones have, in the words of Nicholas Carr, 'made media machines of us all'.[4]

MAKING OUR MARK

The urge to make marks that record events and share stories runs deep; the first known cave paintings were made by Neanderthals more than 60,000 years ago.[5] As hunter-gatherer societies settled and began to farm, life became more complex. Individuals took on specialised roles in specific crafts and trades. Interactions between people expanded, and soon the sum total of useful information greatly exceeded what any individual could carry around inside his or her head. The first written communications that appeared were not examples of free expression but remedies for forgetfulness.

One form of communication in particular began to cause problems: the use of numbers. Our brains are not equipped to process and store numerical data reliably; to keep a tally of trades, rents and taxes and to settle disputes, our ancestors needed reliable transactional records. So the first written words were not epic poems, stories or grand statements of constitution; they were mundane account-ancy documents.

The earliest formal record-keeping systems were established in Sumer, during the fourth millennium BC. These tablets, imprinted with cuneiform scripts, were made by pressing the wedge-shaped tip of a cut reed into a block of wet clay. The British Museum have in their collection one of the oldest known cuneiform tablets – on it, a series of stylised, picture-derived symbols details the wages due to a team of workers. An early example of an employment contract, it indicates that payment should be in the form of beer rations; this was before the invention of money.

With such unambiguous records in place, people who did not know each other could be reassured that each side of a bargain would be ful-filled. With a system to build and validate trust, trading agreements could securely extend into the future. As the American banker J. P. Morgan noted in 1912, trust is critical in all that we do: 'A man I do not trust could not get money from me … that is the fundamental

FIGURE 4.2 Beers for labour, written in clay: a 5,000-year-old clay tablet from Mesopotamia, an early form of an employment contract.

basis of business.' Today, transactional records are as important as ever, precisely because they foster trust. The reputational scoring systems used by online shopping sites like eBay and travel services such as Uber ensure that today's interactions are carried out with an awareness of their implications for tomorrow. By broadening the scope for trustworthy relationships, wealth grows, ideas emerge and progress continues.

While the oral tradition is powerful, its reliability is limited, and meaning inevitably drifts with time and repetition. Gradually, symbols stopped merely representing an object, but also came to stand for its sound. The emergence of true alphabets meant that written words could become flexible stores of meaning, and literacy became something people could obtain with a few years of training rather than decades of hard work. Writing became a way to record, share and learn from the stories, laws and calculations of the past.

Mathematics is the only truly universal language, and it underpins all commercial transactions. Scientists use it to measure the world and to judge the importance of what they find. Before they build, engineers must first use mathematics to calculate and forecast. It is the only language that our machines can understand. We call the numbers that we use today Arabic numerals, although they were actually invented in India during the first millennium AD. A seventh-century mathematician and astronomer called Brahmagupta introduced a symbol and a formal concept for zero – despite its superficial lack of substance, it is the foundation of Cartesian geometry and of calculus. Without these mathematical tools, progress would have ground to a halt long ago.

Mathematicians in the Middle East took their subject in a new direction. The word 'algorithm' comes from a sloppy transliteration of the name of the most influential Persian mathematician, Al-Khwarizmi. The word 'algebra' comes from the title of his ninth-century treatise, *Al-jabr*.[6] Mathematics was now much more than just a way to balance the books; it was a way to probe truths about the world, to construct hypotheses and to test relationships. As Brahmagupta said, 'As the sun eclipses the stars by its brilliancy, so the man of knowledge will eclipse the fame of others ... if he proposes algebraic problems, and still more if he solves them.'[7] This is what computers do today – they solve algebraic problems.

PRINTING AND THE POWER OF REPRODUCTION

A German goldsmith called Johannes Gutenberg took the next great step forward in the 1440s, when he made the first printing press.[8] It was an innovation that demonstrated the power of combining existing tools and techniques to create something new and powerful, which to this day is a route taken by many innovators. Firstly, Gutenberg used his mastery of metalworking to cast individual letters that were absolutely uniform in size and shape, which meant they could be quickly and precisely stacked in neat rows to produce what we now call type. Secondly, he used paper that had been produced in Germany since the fourteenth century. Thirdly, he adapted the chemical formulation of ink to make it capable of being spread on the type. Finally, he used the screw press from winemakers to apply pressure to the paper once it had been laid on the type. And so, in 1455, he produced an edition of the Catholic Bible that was the first book to be reproduced hundreds of times without its contents changing at all. These books were expensive to produce, some of them costing the equivalent of three years' wages for a clerk, but they were much cheaper than manuscripts handwritten on vellum. As with most production, the cost of printed materials kept falling as more were produced.[9] Printing quickly propagated across Europe, and beyond. Within two decades, at least 350 presses were turning out texts, and millions of books were being printed.[10]

Inevitably, there were doubters who bemoaned this new invention. During the late fifteenth century, Venice emerged as the most influential printing centre in the world; in 1473 the Benedictine monk and scribe Filippo de Strata wrote to Venice's doge, complaining about the new scourge of printing: 'They shamelessly print, at a negligible price, material which may, alas, inflame impressionable youths, while a true writer dies of hunger ... Cure (if you will) the plague which is doing away with the laws of all decency, and curb the printers.'[11] The monk's complaint now seems a precursor of modern-day concerns about social media platforms.

Advances in printing and reproduction were not just about sharing words and numbers. Standing in the Print Room of the British Museum with Antony Griffiths, the former Keeper of Prints and Drawings, I am struck by how revolutionary the advent of printed images must have been. Paintings and drawings had previously appeared on the walls of churches and palaces, but almost nowhere else; when the print came along, it liberated the image from these exclusive settings. 'The great thing about the print was that it was a single sheet of paper which carried an image. You didn't need a text; the image was the message and anyone could own it,' explains Griffiths.

The printed picture represented a particularly powerful form of communication at a time when most people could not read. Prints soon became an effective way of spreading propaganda and political messages. A powerful example is Titian's monumental woodcut *The Submersion of Pharaoh's Army in the Red Sea*. This is an audacious print, which he produced in around 1514 by joining twelve individual woodblocks together to create a single image that is more than two metres wide. It depicts the well-known biblical scene from the Book of Exodus of Moses parting the Red Sea to allow the Israelites to escape from Egypt, but there can be no doubt that Titian intended to convey a much more contemporary message. Venice, against all the odds, had just survived an invasion by the powerful armies of the League of Cambrai; it seems that Titian's aim was to show that the Venetians had made the great escape, as their invaders from the north drowned in the Red Sea. In the foreground, a dog defecates in the direction of the floundering army; the message is clear.

Many of the new applications of printmaking did not replicate or substitute existing images, but rather created completely new forms. 'Think of the whole fashion industry,' says Griffiths. 'That couldn't have happened without prints showing people what the fashions of, say, 1685 looked like.' Perhaps more pertinent for this story of progress

FIGURE 4.3 Powerful propaganda through the power of print: Titian's *The Submersion of Pharaoh's Army in the Red Sea* (1514–15). Spot the dog.

was the ability of printing to make exact reproductions of scientific and technical diagrams. This had the effect of standardising scientific and technical knowledge, much of which can only be clearly explained in visual form. This can be seen in some of the earliest technical books about science and engineering. The prints in Agricola's *De re metallica* (1556) and Biringuccio's *De la pirotechnia* (1540), the two seminal books that describe the mining and metalworking techniques of that time, have a vividness and a level of detail that much of the accompanying text cannot match. As Griffiths explains, suddenly 'you could transmit information, and you could all know what you're talking about, because you could all be looking at the same thing, at the same time, in different places'. As a result, science and engineering took a great stride forward.

The advent of an abundance of printed words and images resulted in an explosion of new sources of information and inspiration. Before the arrival of the printed book, scholars had been preoccupied with creating accurate versions of important classical texts from badly copied versions made by generations of scribes. Clearly, the most reliable texts were the oldest ones, and these could now be copied with a printing press. Scholars then had the freedom to look forwards rather than backwards; the ability to circulate concepts, discoveries and news fostered communities of critical thinkers. Rational enquiry advanced through collaboration and competition. During the Renaissance, progress spread through science, engineering, economics, politics and art; new devices, including microscopes, telescopes and barometers accelerated the pace of discovery and challenged more and more of the received wisdom from the ancients. The time of the isolated inventor had passed – progress now came from humanity's steady accumulation of knowledge, which was lubricated and fuelled by the ability to connect people and ideas. Over the course of four centuries, printed texts and images had spread to almost every corner of the world. The next great step came when innovators learned how to encode information and convey it electrically, over long distances and at previously unimaginable speeds.

'WHAT HATH GOD WROUGHT?' THE RISE OF ELECTRONIC COMMUNICATIONS

On 24 May 1844, Samuel Morse used his eponymous code to tap out a message. It was transmitted for forty miles along a slender copper wire, from the Old Supreme Court Chamber in Washington, DC to Mount Clare Station in Baltimore. Decoded, the words of the first long-distance electric telegraph asked 'What hath God wrought?', a question that uncannily forewarned of the disruption that such communications would ultimately cause. Very soon, a global telecommunications network began to be built and, by 1858, a telegraph cable had been laid across the floor of the Atlantic Ocean. US President Buchanan's first message was a bombastic assertion of his belief that transcontinental telegraphy was 'an instrument destined by Divine Providence to diffuse religion, civilisation, liberty, and law throughout the world'.[12]

For the first time, cities and towns across the planet were physically connected by information-carrying wires. In an imperial age, the 'highway of thought' allowed nations to keep better control of their subjects, spread propaganda and rule with greater speed and decisiveness. The news business took on a whole new scale and immediacy as wire agencies relayed their messages around the world.

Business relied on the telegram and its later incarnation, the telex, until deep in the twentieth century. During the late 1970s when I worked at BP's London headquarters as assistant petroleum engineer for the North Sea region, I vividly remember waiting for the daily incoming telexes from the rigs that were drilling in the region. They were printed on green paper, written in an arcane shorthand, and when the drilling results were commercially sensitive, they were coded. Decoding was done by hand; we were lagging behind the achievements of the cryptographers of the Second World War.

Soon after the telegraph came the telephone, which was first invented in 1854 by Italian-born engineer Antonio Meucci and then improved and commercialised by Alexander Graham Bell. Its ability to allow natural conversations in real time, across geographical distances, was revolutionary. At first, many people were wary because they were concerned about their privacy; they suspected that strangers could listen in to their private conversations. For some, like my mother, who grew up in Eastern Europe during the 1930s and 1940s, these concerns were understandable. Other concerns about increasing the pace of life and being continuously disturbed vanished, as people began to trust and rely upon the new invention. However, it was expensive to use and people kept phone calls brief since they were charged according to their duration – when I worked in California during the 1970s, I remember calls home to England costing $3 per minute for the first three minutes (about $20 per minute in today's money). It is only recently that Voice over Internet Protocol technology has allowed us to make the same calls for no obvious cost.[13]

The next significant advance was the radio. When Italian electrical engineer Guglielmo Marconi first demonstrated a primitive radio receiver in a public lecture in 1897, he was immediately compared to the illusionist Harry Houdini, but, undeterred, he continued to defy expectations. Physicists had shown that electromagnetic waves, like beams of light, do not bend. Most were convinced that the curvature of the Earth would make transatlantic radio transmission impossible, but Marconi tried anyway. His experiment worked; as the sceptical physicists learned, the upper atmosphere could reflect radio waves. Marconi described the exciting moment, in 1901, when he waited for the first transatlantic signal: 'I was at last on the point of putting the correctness of all my beliefs to test. The answer came at 12:30 when I heard, faintly but distinctly, "pip-pip-pip" … I knew then that all my anticipations had been justified.'

The first live news bulletin was broadcast on 31 August 1920, in Detroit. But, predictably enough, radio was soon cast as the next threat to our well-being. In 1936, one magazine opined that children were developing 'the habit of dividing attention between the hum-drum preparation of their school assignments and the compelling excitement of the loudspeaker'. By the end of his career, even Marconi

FIGURE 4.4 Breaking radio silence: Marconi's ingenious radio broadcast transmitter, Britain's first (Chelmsford, 1919).

was questioning the value of his invention, asking 'Have I done the world good, or have I added a menace?' His invention had opened the door to today's highly interconnected world of mobile telephony, Wi-Fi and Bluetooth.

The new channels of telecommunication, which appeared in the late nineteenth and early twentieth centuries, lowered the barriers to the flow of information from one place to another. In doing so, commerce and knowledge were able to spread more rapidly. Communication meant more innovation, since competitors could rapidly find out about new inventions and copy and improve them. Governments and media organisations could deliver their messages, whether true or false, directly to the population. Corporations could have a dialogue with their staff to find out what they wanted. But this was not the start of open communication, since only the most powerful and influential had access to the technology to send their messages directly. That is why radio and TV stations were always an early target of military coups.

The Internet, powered by the World Wide Web and first made accessible to a global audience by development of the Web browser, changed all that. The Internet is an engineered network of vast scale and complexity. In 2017, more than half the world's population used it,[14] and around 80 per cent of these users owned a smartphone, which allowed them to access the Internet whenever and wherever they wished. Through social media and collaborative efforts such as Wikipedia, every user's actions, ideas and opinions can ripple through the whole network and anyone, anywhere can, in principle if not always in practice,[15] enter into a dialogue with anyone else. Making the Internet work has been an audacious engineering achievement.

The World Wide Web is, as Nigel Shadbolt, Professor of Computer Science at Oxford University, describes it, 'a beautifully engineered minimal set of protocols, with just enough power to enable documents to be connected across the planet'. The simplicity of its design reduces the opportunities for failure, and therefore makes Web-based communication highly reliable. Together, the Internet and World Wide

SHARE OF INDIVIDUALS USING THE INTERNET

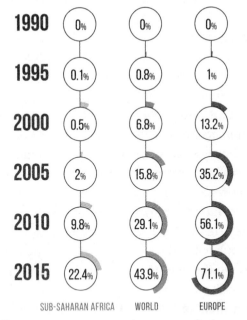

	SUB-SAHARAN AFRICA	WORLD	EUROPE
1990	0%	0%	0%
1995	0.1%	0.8%	1%
2000	0.5%	6.8%	13.2%
2005	2%	15.8%	35.2%
2010	9.8%	29.1%	56.1%
2015	22.4%	43.9%	71.1%

FIGURE 4.5 From zero to 44 per cent in 25 years: the increasing proportion of people using the Internet.

Web link us all together; it shows the potential of devising clear, robust principles and then applying them on a massive scale.

THE INTERNET WEAVES ITS WEB

The Internet first existed as the ARPAnet, which was developed by the US Department of Defense in the 1960s. It was designed to increase access to the small number of powerful computers in institutions scattered around the country. To make it work, computer engineers needed to make sure that data could not get lost as it made its way across an unreliable telecommunications network – telephone conversations can accommodate a degree of signal loss, but computer programs cannot.

The pivotal innovation was to break data up into small packets, label each packet with its ultimate address, and then allow it to plot its own course, calculating its next move according to the congestion it meets, before being accurately reassembled with other packets at its destination. This process, known as 'packet switching', is the key to the success of the Internet. Its co-inventor, Vint Cerf, 'knew this was extremely powerful technology, [since it] let any computer on any network communicate with any other computer on any other network'. Importantly, Internet protocols make no assumptions about the systems that carry the information – as optical fibres, high-speed satellite communications and mobile Internet came along, they were incorporated seamlessly into the network. As Cerf puts it, 'the system has this incredibly future-proofed appetite for new underlying transmission technology'. So, as people wanted to share more texts, images and then videos, the bandwidth of the network has expanded seamlessly. Importantly, the Internet is completely agnostic as to the applications that use the information that it carries. One of the first, and most influential, was email.

Ray Tomlinson, a young engineer working on the ARPAnet, sent the world's first email in 1972. His first message, with the now-iconic '@' symbol, was sent between two terminals in the same room – nowadays, some 270 billion emails, around forty for every member of the human race, go back and forth across the network every day,[16] though more than half of these are unsolicited spam.[17] Technology entrepreneur Suranga Chandratillake argues that email is a 'democratising innovation', because it allows anyone to transmit their ideas to whole communities, wherever they are on earth. Chandratillake thinks of many of today's communications innovations, from Facebook messages and video calls, to shared 3D virtual environments, as 'derivatives of that core object, the email'.

The transition from ARPAnet to Internet, and the explosive growth that followed, was catalysed by Tim Berners-Lee's creation of the World Wide Web. The Web made it easy for anyone with a basic grasp of computing to create web pages and make them accessible to the world. Berners-Lee's original scheme was judged by some to be too simple. The first academic memo that he wrote describing his invention was almost entirely ignored for over a year until May 1990, when his supervisor at CERN, Mike Sendall, encouraged him to test his system.[18]

Since World Wide Web software was introduced to the outside world in August 1991, many aspects of contemporary life have changed. On the phone from his office at MIT, Berners-Lee stresses that the Web is an engineered system that derives its power from the ease with which it allows people to connect. 'You can call it all engineering,' he says, 'but half of it's social, and only half of it is technical.' Under Berners-Lee's guidance, Web technology has remained free and open, a common resource for humanity. This is the source of its energy, but also of its vulnerability. As Daniel Weitzner, Berners-Lee's colleague at MIT wrote, 'by designing the Web to represent all of humanity, Tim posed a challenge for all of us. As a decentralised medium, no single authority can filter out the bad and save the good.'[19]

Like telegraphy had been before it, the Web was quickly seized upon by optimists and idealists. In 1996, the late 'cyberlibertarian' John Perry Barlow declared that the online world was a parallel realm that was superior to, and unsullied by, the squabbling and inequity of terrestrial governance, writing bombastically that 'we will create a civilisation of the Mind in Cyberspace. May it be more humane and fair than the world your governments have made.'[20] Naïve as this opinion may now sound, establishment figures also caught on to the excitement surrounding the Web's arrival. In an internal Microsoft memo in March 1995,[21] Bill Gates described how the Internet could create a new world order. 'The Internet is a tidal wave,' he concluded simply. 'It changes the rules.' MIT professor Nicholas Negroponte told a conference in 1997 that he had 'never seen people miss the scale of what's going on as badly as they are doing it now'. Children growing up in the 1990s, Negroponte declared, were 'not going to know what nationalism is'. He dared to believe that better communication could eventually lead to world peace.[22]

Some were not convinced. 'Baloney,' wrote the US astronomer Clifford Stoll in 1995. 'The truth is no online database will replace your daily newspaper, no CD-ROM can take the place of a competent teacher and no computer network will change the way government works.'[23] Stoll believed that the new medium would not catch on at all, but rather humbly corrected himself in 2010. The Nobel Prize-winning physicist Arno Penzias made a more accurate prediction – I remember hearing him say that the very openness of the Internet would become its Achilles heel. He feared that it would not make the world more interconnected but rather that it would ultimately result in balkanisation, xenophobia and isolation. Swept along by the optimism of the day, I didn't agree with this assessment at the time – how could people fail to connect and build a richer, more inclusive civilisation?

FAKE NEWS AND FILTER BUBBLES: THE ONLINE ABYSS

In today's online world, some of Penzias' bleak predictions have materialised as, faced with seemingly endless choices, people instinctively turn to the familiar. The architecture of popular Internet services magnifies this tendency. The algorithms behind Google's personalised search results, for example, are designed to learn what we like and to deliver the most relevant content. These services do their job well, but they can also shield us from opinions that challenge our view of the world. Cerf now has the extraordinary job title of Chief Internet Evangelist at Google, but even he is concerned about this problem. 'What is truly disturbing ... is that it allows you to create your own universe and to ignore all other information, except that which you find acceptable.'

Writer and thinker Eli Pariser gave this phenomenon a memorable name: The Filter Bubble. He writes that, 'A world constructed from the familiar is a world in which there's nothing to learn ... [since there is] invisible auto-propaganda, indoctrinating us with our own ideas.'[24] That may even, rather dangerously, apply to populist politicians who are encouraged to believe their own propaganda. The philosopher Francis Bacon described this same behaviour 400 years ago, when he wrote that 'Everyone ... has a cave or den of his own, which refracts and discolours the light of nature.'[25]

The tendency for algorithms to curate news and opinion in ways that reinforce partial and bigoted viewpoints will undo progress towards more inclusive and tolerant societies, and is particularly dangerous when content is fabricated or deliberately misleading. Reflecting on the Democratic Party's defeat in the 2016 US presidential election, former President Barack Obama bemoaned the fact that we live in 'an age where there's so much active misinformation ... If we can't discriminate between serious arguments and propaganda, then we have problems.' Nigel Shadbolt is similarly alarmed by the politicisation of social media – the great scale and reach of social networks mean that political messages, whether true or false, can be targeted at very specific groups of people: 'You have never been able to get to the swing votes in quite the same way. You can shift "don't knows" by an explosion of information on a particular topic,' he says. One analysis concluded that fake news influenced the American electorate in 2016 as much as reports from reputable news agencies.[26] A detailed study of all the items posted to Twitter over more than a decade showed that 'fake news' and false rumours consistently spread more quickly, reaching many more people than truthful and accurate stories.[27] Our attention is most easily captured by information that is both novel and threatening. As the political scientist Rebekah Tromble commented, 'It's all too easy to create both [of these features] when you're not bound by the limitations of reality. So people can exploit the interaction of human psychology and the design of these networks in powerful ways.'[28] This unfortunate quirk of our psyches goes some way to explaining why the bad news always outsells the good and, by extension, why so many people are sceptical of the idea that humankind is making steady progress towards a better world. The propensity for emotionally resonant and threatening messages to be amplified online is contributing to the alarming rise of authoritarian and populist politics in many parts of the world. These movements knowingly exploit the fact that, in the battle for our attention, menace and negativity triumph over positivity, reason and informed debate. The problem of 'fake news' is not a new one – people have long been using the communication media of the day to stretch the truth and promote falsehood.[29] However, the philosopher Onora O'Neill thinks that what is new is the power we have to inject our views into a network that spreads them far and wide: 'If I say something idiotic, I make an ass of myself, but if a newspaper does, lots of people may act on it,' she says. 'What are we doing to democracy with these technologies that circulate the agreeable at the expense of the accurate?'

Deciding what in our media is true is never a task to take lightly. In Renaissance Venice, would-be publishers had to apply for a licence to print books, and many who were judged by the authorities to have spread falsehoods were marched over the Bridge of Sighs to the cells for what was often a trip from which they would never return. Nobody now is calling for such a centralised and draconian system of truth arbitration, but the challenge of providing systems that help people to judge veracity must be addressed. If it is not, Cerf thinks that 'the online environment will become so toxic that people won't want to use it'. Berners-Lee, however, is optimistic that redesigning Web services will make people behave better online. 'It's not a question of building a supervisory, controlling truth police on top of our social networks,' he explains, 'it's much more a question of defining the rules of play.' He describes how a football match can become very fierce, which is fine so long as everyone understands the rules and that they will be sent off if they break them. The challenge, then, is to devise rules that encourage open and constructive interactions across the Web. 'Rather than policing people who put badness on Twitter, suppose we change Twitter's algorithm, so that people are naturally more constructive,' suggests a very optimistic Berners-Lee. The kind of system that he envisages would reward socially beneficial messages and would have the intelligence to query content that appears to be inflammatory or offensive, before it is published. These systems could ask, for example: 'Well, you criticised it, but could you please provide an alternative to their proposal?'

Rewriting the rules of online engagement could help, and it is also possible that we can develop engineered systems to differentiate between the fake and the true by comparing all sources with a statistical analysis about which sources are verifiably accurate. However, I am certain that we will be unable to dispense with our own judgement, and Cerf takes a similar view. I meet with him at the Folger Shakespeare Library in Washington, DC. Pointing to the great plays and books that surround us, he exclaims that 'We learn, once again, why we still read Shakespeare's plays; people haven't changed in the last 400 years.' For Cerf, the constructive and destructive uses of the Internet are inevitable facets of the human condition. He does not think we can rely solely on technology to filter out all the bad intentions of our fellow humans. Just as people did in Shakespeare's time, we have a responsibility to 'pay attention to what we're reading and seeing and hearing, and to try to assess whether or not it's worthy of our attention'.

TRADING OUR PRIVACY

Onora O'Neill is adamant that most of us are too trusting of the technology companies that provide our media and broker our social connections across the network. A recent report points to decreasing levels of trust in business, government, NGOs, the media and many other key institutions. However, in all but one of the twenty-eight countries surveyed, respondents claimed to trust the technology industry.[30] This predisposition to trust is borne out in the way we behave online. When installing new software updates, most of us happily 'agree' to page after page of dense legal language. O'Neill wonders whether our attitudes to online transactions even qualify as trust at all: 'It is tempting to say that [the online world] is a trust-free zone, with a lot of credulity on the surface.' She points out that people are 'being asked to place and refuse trust in complex systems, rather than in something they understand pretty well'. As a result, she asserts, it is much harder for us to trust intelligently.

In many cases, we casually consent to sharing a wide range of personal data, from our whereabouts, to the contents of our emails and our online search histories. In many of these decisions, con-venience seems to trump privacy and security. A recent experiment asked students to make choices based on how easy it was to access information rather than on the privacy implications of their actions – the results suggested that, regardless of how concerned they claimed to be about privacy, ease always won.[31] Alarmingly, a small incentive, such as a free slice of pizza, seemed sufficient to overturn concerns about privacy.

Trust is essential for society to function, and too much or too little can cause civilisation to break down. Trust too willingly and you are open to exploitation, but if we cannot gauge trustworthiness, and therefore withdraw our trust, progress grinds to a halt. Early in 2018, it emerged that the personal data of around 87 million Facebook users had been acquired by the political consulting firm Cambridge Analytica and used to influence voters in the 2016 US presidential election and the UK's referendum on the membership of the European Union; these revelations prompted a backlash against Facebook.[32] Ed Williams, CEO of public relations company Edelman in the UK and Ireland, is critical of the way in which Facebook CEO Mark Zuckerberg explained his company's actions, and says, 'I think it's particularly clear that their lack of immediate engagement, and the initial tone deaf nature of their reaction, risked a blunt and unsophisticated regulatory response.' Governments in the US, the UK and elsewhere are scrutinising Facebook's activities.

Vint Cerf warns that we must not confuse privacy with anonymity when thinking about better ways to conduct our online interactions. 'It's my belief that privacy is a kind of anomalous condition,' he says. He goes on to describe his experience of living in a small German village in the 1960s, when all phone calls, telegrams and letters were organised by the postmaster who ran the village's single post office. 'Everyone knew what everybody else was doing. There wasn't really much privacy at all.' As cities grew, Cerf thinks people felt their privacy was protected because nobody knew who they were. 'Of course this wasn't true either,' he continues, pointing out that online environments have made it ever easier to unearth information about other people. 'Privacy is very hard to come by,' concludes Cerf, and 'we need to be conscious of that and work at maintaining privacy.'

Andy Hopper believes that trust in data is essential, saying that 'knowing the provenance of data should be a key principle of all computer systems'. Computer scientists in his laboratory at Cambridge University are devising ways to keep track of the source of data and everything that happens to it over its lifetime automatically – if this works, misrepresentation and forgery could be identified and weeded out. When necessary, data could be tracked and controlled, even when shared with private corporations or public institutions. Hopper believes that everyone should be able to see who holds their data, how they are using it and whether they are monetising it. Moreover, everyone needs guarantees that their data is held securely and should have the right to delete it should they choose to do so. Berners-Lee thinks that the Web, which started out as a decentralised and open system, has become too subject to centralised control. 'It has always been a goal of mine to keep the Web private,' he says. 'That battle has become much bigger because governments realised what a huge benefit there is to essentially being able to spy on citizens.' He also recognises that the huge scale of social networks like Facebook and Google allows them to provide high quality services, but fears that they now have 'complete control over the user's life ... by controlling their newsfeed, and controlling the way they see the world'. Berners-Lee foresees 'big battles ahead about who software is really reporting to'. For example, when the software that he uses to do his accounting tried to sell him insurance, he felt it had crossed a line. He says, incredulously, 'I've paid money for you, you're a piece of software. Philosophically, you work for me – you don't sell me insurance. They're called your "personal assistants", but who does Siri work for? Who does Alexa work for?'

At MIT, Berners-Lee is designing systems that he hopes will reverse this dynamic and give people more control over their personal data,[33] by not putting it on the cloud but leaving it under their local control; a digital service would interact with their data only when specifically requested. If that can be achieved, everyone stands to gain. As Berners-Lee says, 'The value that other people get out of my data is, I posit, miniscule compared to the value I get out of it. Nobody is more interested in what I'm buying, what I'm interested in and my health, than me.' Nigel Shadbolt comes at the issue from a different angle, believing that 'the benefit that accrues from my data being amalgamated with many other people's to provide insights about general conditions is an inarguable thing'. He thinks the data owner should be empowered to permit this to occur and that 'it's very odd that the narrative isn't around empowerment. It's about the risks'. Concerns about privacy are real, but solutions must be carefully designed so that they do not limit the huge potential of 'big data' to provide unique insights. Mark Walport believes that often 'privacy is used as an excuse' for not sharing medical information, despite the fact that 'there's a lot of evidence that when you are ill you want to share your data as much as you possibly can'. Big data has the potential to revolutionise medical diagnosis and disease prevention. As John Bell, Professor of Medicine at Oxford University, put it to me, 'genetics is the perfect example [of a situation] where there's a general responsibility for everybody to share their data so that we can advance the knowledge, so that we all benefit'. There is a risk that over-zealous attempts to safeguard privacy could destroy that benefit.

Concerns about privacy are not, however, universally shared. The Chinese government, for example, is in the process of introducing a new 'social credit system' that, when fully implemented, will monitor personal data from many different sources, including people's purchasing habits, credit history, social media contacts and content, and media consumption. Using an algorithm, it will then integrate all this data, in order to grant every Chinese citizen a single, publicly visible rating that reflects their trustworthiness.[34] People with a high rating will be rewarded with special privileges, such as access to loans, but those with low ratings will find it harder to access credit, employment and travel opportunities. This may make it easier to decide the trustworthiness of a person, but it could also be used as an integrated, government-sponsored incentive system to make people conform to a specific model of citizenship. However, it is already the case that people in many other countries are rated by similar systems, though this is done on a more piecemeal basis and not in a publicly visible way. Cryptographer and computer security expert

Bruce Schneier believes there is a way to get the advantages of rating systems without having to accept their drawbacks.[35] Firstly, the workings of the algorithms must be accessible and explained to the public in comprehensible terms, so that everyone understands how their behaviour affects their rating. Secondly, making prejudicial decisions on the basis of algorithmic judgements must be illegal, and for it to be enforceable, the creators of the algorithms must be subject to independent regulatory scrutiny. Finally, there must be channels through which anyone can challenge judgements that are made. As Schneier notes, 'we need the ability to clear our names' – people must be given the opportunity to explain the nuances and complexities of life that may have forced them to make decisions that negatively affected their ratings.

Society has no need to overreact to the challenges presented by our networked world. Despite the concerns of the sixteenth-century scholar Conrad Gessner, movable type did not lead to a 'confusing and harmful abundance of books'.[36] Worrying claims of the declining attention spans and screen addictions of today's teenagers are largely unsupported by good evidence[37] – in fact, millennials may be reading more books than their elders.[38] In a recent survey of UK secondary school pupils, 60 per cent of respondents claimed they 'wouldn't mind if social media had never been invented' and an even greater number reported that they have taken breaks from using social media and plan to do so again in the future.[39] It appears that younger generations may have a far better understanding of the social impact of network technologies than older generations. 'Could the "digital detox" movement become a trend?' asks Ed Williams. And, at a time when so-called 'post-truth' content is rife and free, paid subscriptions to well-respected media outlets are rising fast, with new fact-checking organisations also flourishing.[40] Even more importantly, millions of people around the globe are connecting with other people and ideas for the first time. This is a phenomenon that poorly conceived regulation must never be allowed to damage. As Berners-Lee puts it, any changes to the network's function must not damage 'the right for two people to communicate across the Internet'.

In a reversal to the assumed trajectory of international development, there are now more mobile phones in Africa than in Europe,[41] and they are driving economic growth and stimulating innovation in unanticipated ways. Over the last decade, the same technology that is used to send simple text messages has been co-opted and re-engineered to transform it into a complete financial services platform. The first and most successful of these mobile money schemes, which was launched in Kenya in 2007, is called M-Pesa. Today 25 million Kenyans use the service, which moves the equivalent in value of half the nation's GDP every year.[42] The late Calestous Juma of Harvard University described the local needs that drove this mobile money revolution – workers in Kenyan towns and cities urgently needed a way to send money to relatives in isolated rural districts, who before M-Pesa had to wait weeks or months for that crucial income. Equally necessary was a reliable way to grant and repay loans in microfinance schemes; these small sums, often paid directly to women, can provide the means for families to build their own way out of poverty. 'What I find interesting is that ... what started as a consumer product became an engineering platform,' explained Juma. Schemes like M-Pesa are now spreading across the globe and evolving to provide new services. They can be used to purchase insurance and credit, to receive wages and to pay taxes. Crucially, transactions are secure, traceable and resistant to corruption, which are features that build trust. Juma called this powerful process by which communities absorb technological advances and adapt them to their own circumstances 'inclusive engineering'. 'If more people can have access [to new technology] ... they will come up with all sorts of industries that have not even been imagined yet.' With the spread of mobile Internet and the rise of the smartphone, many inventive new services are already emerging.[43]

THE COMPATIBILITY OF HUMANS AND MACHINES
At MIT Professor Breazeal builds and conducts research with socially attuned, empathetic robots. She is passionate and convincing in her belief that robotics and artificial intelligence can be made to serve

the public good. 'The dream is for it to become a ubiquitous technology,' says Breazeal. As well as a computer on every desk, she wants to see a social robot in every home, classroom and hospital. At the heart of her mission is an aim to use robotics and artificial intelligence to deliver high quality, personalised support 'to everyone, but especially to the people who are the most vulnerable and in need'. For the best outcomes, daily interactions at home with personal tutors, health coaches and other care professionals are usually necessary. This is, however, very expensive and not scalable to meet the growing demand. But with social robots, 'for the first time we have a way to deliver truly scalable, affordable, high-calibre, personalised intervention to everyone who needs it', says Breazeal.

Kenichi Yoshida of SoftBank Robotics describes how hundreds of hospitals in Japan are already using his company's humanoid robot, which is called Pepper. He explains that rapid advances in perception and understanding have allowed these robots, which are most often used to conduct pre-screening in waiting rooms, to communicate with patients in increasingly sophisticated ways. He describes how Pepper might, for example, engage patients with a questionnaire, analyse the results and then announce, 'OK, you've got a risk of bone density disease. Why don't you discuss it with your doctor?' Another application, which Yoshida calls 'a kind of brain training', converses with people with dementia, to keep them mentally active and responsive. As robots become more socially attuned and communicative, leading surgeon Ara Darzi tells me that they could help combat social isolation, which he called 'the biggest cause of disease [in our expanding elderly populations], causing depression, anxiety and cardiovascular disease. There are early signs that interactions with machines fill that gap.' All of these uses of robots can help relieve medical staff of some of their workload and improve the standard of care.

Breazeal rejects the idea that interaction with social robots could ever replace human communication and relationships. 'People, concerned about the possible dystopia of success gone too far, often ask me about what will happen when people actually prefer their relationships with social robots instead of human relationships,' says Breazeal, 'but I have studied this long enough and I am not worried about that at all. We, as human beings, are capable of many, many, many different kinds of relationships, so I don't see robots replacing human relationships, but adding something new and complementary.' We are a profoundly social species, and need human relationships to thrive. 'We need to feel we belong, are valued and can contribute to those who matter to us,' Breazeal adds. She thinks that social robots could actually remind us of the importance of physical interaction. She says that, according to psychological research, 60 to 80 per cent of human communication is non-verbal and based on tone of voice.[44] Robots that can understand and express these important cues will engage us and add important new dimensions to our interactions with computers and machines. 'At least [when you relate to a robot]

FIGURE 4.6 'Call me Doctor Pepper. I am a humanoid robot and can talk with patients for pre-screening.'

you are keeping your head up and not staring at a screen,' she adds.[45] For Breazeal to realise her dream of a world populated by responsive technologies that are more attuned to our needs, two crucial innovations are missing. Firstly, we need more reliable ways to communicate with our machines and secondly, they need to get much better at understanding precisely what we are saying.

I meet Rahim Tafazolli, Professor of Electronic Engineering at Surrey University, in a restaurant in West London to talk about how 5G, the fifth generation of mobile Internet network, will lead to an era of 'mass connectivity'. Over a meal of fesenjan, a stew of chicken flavoured with walnuts and pomegranate that reminds me of my childhood in Iran, Tafazolli explains how 5G will allow smartphone users to connect to the Internet ten times faster than they can today. 'The second and, in my opinion, most transformative aspect of 5G is automation,' he adds.

At the 5G Innovation Centre in Guildford, Tafazolli is helping to design the infrastructure that will help the widely advertised 'Internet of Things', a catch-all term that describes all Internet-connected devices, deliver on its promises. 5G networks will connect millions of machines to the Internet, and to each other, with great speed, large geographical coverage, low latency and, very importantly, high reliability. Reduced latency is crucial for some applications. 'Robots don't understand delay; they can't wait until they receive commands,' Tafazolli explains. Guaranteed low-latency, high-bandwidth and ultra-reliable 5G networks will help make autonomous cars and drones as safe and efficient as possible. Tafazolli describes 5G as 'the connectivity that brings automation to manufacturing, automation to health care and automation to make the home smarter, cities smarter and the planet smarter'. Because 5G networks will allow more devices to take on 'mundane, repetitive' tasks, Tafazolli is confident that it will be 'transformative of the digital economy, the connected society. It will affect most aspects of our lives'.

Ryan Ding, Executive Director and President of Products and Solutions at the technology company Huawei, acknowledges the great potential of 5G connections, but also cautions that many parts of the world must improve more basic aspects of their digital infra-structure if they are to reap the benefits of the connected world. He shares the translation of a Chinese phrase that he feels underpins the country's dramatic economic resurgence in the last forty years: 'If we build a new road, all the villages and towns along this road become rich.' China has appropriated this logic for the digital age – Ding describes how the city of Shenzhen alone has more than two to three times as many 4G base stations than either the whole of Germany or the UK. 'From my point of view, the whole of Europe is an emer-ging market for 4G coverage in the telecommunication industry,' he concludes emphatically. However, even if we have the infrastructure that we need to connect to machines seamlessly and immediately, we will be able to exploit their full potential only if they can understand what we are saying.

In the 1968 film *2001: A Space Odyssey*, the computer system HAL 9000 stars as an omnipresent, intelligent and highly communicative machine. 'I remember being very affected by going to see the film as a young boy,' Nigel Shadbolt tells me. During the 1990s, he began to see the limitations of systems that 'tended to exist locked in computers on desks' and started to see the Internet as a way to leverage the power of network effects, to create more intelligent computing structures. His dream is to transform the Web from a passive conduit of data into a system where machines can understand the information with which they are dealing. Over the last two decades, Shadbolt and Berners-Lee have been spearheading attempts to build and promote the 'Semantic Web'. This is starting to happen. 'There has been a quiet revolution,' says Berners-Lee; today's 'Web of documents', most of which are only interpretable by humans, is gradually becoming a 'Web of data', all of which will be intelligible to computers.

At Google's campus in Mountain View, California, I meet Hadar Shemtov, a Director of Engineering who wants to make it easy for us to have fluent and sophisticated conversations with our computers. Shemtov describes conversation as 'a very complex game' because, 'when I start a sentence, you already have a model of what I'm going to say'. To be able to succeed in this game, computers will have to understand the content of what we are saying, but also learn to predict what we are about to say. Algorithms are handling language with more confidence all the time. One of Shemtov's colleagues, Charina Chou, described how a reporter for the *New York Times* recently fed a text by Ernest Hemingway into the company's translation algorithm, translated it into Japanese, and then back into English. 'They're not the same but the meaning is there and that style is still there. The translation system is learning everything it can. It's like a toddler.'[46]

Speaking to computers in a normal way is the immediate challenge. Shemtov regards this as a crucial change to the clumsy ways in which we have been communicating with our devices until now.[47] Over my lifetime, my interactions with computers have progressed from punched cards, to keyboard interfaces, to touchscreens and now to voice control. With each shift, the ease and precision with which we can make our intentions clear has grown. Eventually, the neural interfaces being pioneered at DARPA and elsewhere may take these connections to yet another level. But, in the meantime, we will see the maturation of visual interfaces, which are already lending new dimensions to the way we connect with our devices, and with one another.

A TRILLION DIGITAL IMAGES

For several months every year I live in Venice. My home is on the top floor of an old palace and overlooks the Grand Canal. Many of my friends live in San Marco, on the Canal's opposite side. The walk between these two districts is short, but it involves crossing the Accademia Bridge, which is always busy and, at the height of the tourist season, almost impassable. Tourists want to stand on the crest of the bridge to take photos of the canal and the old palaces stretching out towards the bell tower of St Mark's Basilica, one of the most famous views of Venice. Today, these photos are generally taken on smartphones, and many of them are 'selfies'. Helpfully, vendors on the bridge can sell you a 'selfie stick'. As a resident, I could be annoyed by the crowds and vendors, but somehow I feel privileged that in most of the selfies there is a glimpse of my palazzo, its rooftop, its garden and sculptures. All of this is made possible by digital imaging and CMOS technology – if printmaking liberated the two-dimensional picture from the control of the rich and powerful, image sensors have saturated our world with abundant visual imagery.

The CMOS active pixel sensor is a masterpiece of miniaturisation. It was invented by Eric Fossum in 1993 at NASA's Jet Propulsion Laboratory in California. His original brief was to reduce the size and weight of the cameras carried by NASA's spacecraft. Those used at the time were based on the charge-coupled device (CCD) invented by George Smith in 1969. It was Steve Sasson of Eastman Kodak and Michael Tompsett at Bell Labs who pioneered the use of CCDs to build the first self-contained digital cameras; a 1973 cover of *Electronics* magazine displays the first digital colour photograph – a grainy portrait of Tompsett's wife, 'the face that launched a trillion digital images'.[48] CCD cameras were powerful imaging tools, but the ones used by NASA at the time weighed up to fifty-seven kilograms. By creating a 'camera on a chip', Fossum reduced the volume of the camera by a factor of one thousand. From the start, he had a strong sense of the potential of such compact and powerful digital image sensors,[49] but he was frustrated by how slowly his new technology was picked up by industry. Eventually, Fossum and his then wife set up a small company to commercialise the invention.[50] Things went well and, by 2000, the first camera phone was on sale in Japan.

Smith, Tompsett, Fossum and Nobukazu Teranishi (who invented the pinned photodiode in the 1980s, reducing the size and latency of light-capturing pixels in CCDs and then CMOS sensors) were in 2016 awarded the Queen Elizabeth Prize for Engineering for their innovations, over four generations, which forever changed the way we communicate. This goes far beyond taking selfies and sharing amusing videos of cats – as Calestous Juma pointed out, digital imaging can be particularly transformative in emerging economies, in Africa and beyond. With a smartphone camera, doctors can diagnose patients in remote locations without needing to bring them into a clinic. The cameras are even sufficiently sensitive to measure heart rate remotely.

The devices that we use to record and share the images that we collect by the million are starting to understand the content of what they see. For example, my iPhone unlocks itself when it recognises my face, and social media services now automatically identify and tag people in photographs. These features are convenient, but as robots and computers learn to recognise our gestures and read our emotions we will open up more possibilities. Fossum and Teranishi both worry that some of these developments could deliver new threats to our safety, privacy and freedom; Teranishi argues that facial recognition is a form of biometric identification that works at a distance, without the need to gain our consent, and goes so far as to wonder whether it presents a new threat to democracy. 'There are many, many security cameras everywhere,' he observes, 'and we have the technology to recognise people by face authentication, posture authentication, walk posture authentication.' Ryan Ding points out that these systems can keep us safer by, for example, identifying criminal suspects and terrorists, or by recognising dangerous behaviour. He describes how in the city of Shenzhen in China, artificial intelligence software analyses a network of 1.3 million security cameras, which allows the police to monitor the whole city of over 20 million people from one central office. These systems may do their job very effectively, but they could also precipitate the end of public anonymity. Teranishi worries that in places where 'authorities can monitor everyone', people will change how they act and express themselves. Furthermore, images of faces can reveal much more than names and identities. A recent algorithm developed at Stanford University can correctly distinguish between pictures of gay and straight men with an accuracy of over 80 per cent.[51] The use of this software in places where homosexuality is illegal is a disquieting prospect.

The potential to misuse technology is nothing new. 'It's probably been around since the Stone Age,' observes Fossum. However, what is new is the rate at which communications technologies are changing and reaching ever deeper into our lives and into our psyches.[52] These systems have eliminated the barriers to connectivity; not only can people extract information freely from networks, they can readily inject information back in. Compassionate and innovative ideas can bloom, but destructive ideas can also be shared effortlessly. Fossum is concerned about what this means for the inventors of new enabling technologies – do they have a responsibility to try to foresee and pre-empt negative applications of their tools? 'How do we stop ourselves from opening Pandora's box?' he asks.

Engineering innovations give us access to a rapidly expanding array of ways to connect with each other, wherever and whenever we wish. This fuels our creativity and expands our collective intelligence. As ever, promise and peril are two sides of the same coin, and connectivity also allows the spread of falsehood and potent propaganda. As history shows, little can be done to stop this, though we must strive to balance the false with the true by encouraging the public to access more than one source of information. It seems likely that trust in sources will continue to weaken over the years to come, and I think it is unlikely that algorithms will soon be capable of surveying all sources and making reliable judgements about what is most likely to be 'true'.

There will be an increasing number of data protection and privacy laws enacted around the world. For example, the EU has enacted regulations designed to ensure that technology develops in a way that conforms to what it feels are European ethical values. But it will not stop there; India, for example, is another country that claims sovereignty over data. Regulation must be implemented sensitively and rationally, so as not to prevent big data from being used to improve health care and damaging the ability of people to communicate by fracturing the World Wide Web. In any case, governments will still need to use personal data to improve law enforcement and the provision of public services such as transport.

It is sad but true that, while our privacy is precious, we have long since passed the point where anyone can realistically expect all they do or say, either on- or off-line, to remain private. We have got here because convenience has led us to rely on increasingly inscrutable complex systems, without much regard to the consequences for our privacy. We have become too casual about it; we find it too easy to agree to waive any protection and only protest at an abuse when it is already too late.

Increased connectivity is an aspect of progress that is replete with trade-offs and where the costs and benefits are often hard to separate. In the near future, our ability to communicate will be magnified further, for better or for worse. Soon, we may only have to *think* and our machines will hear our thoughts and reorder the world around us. We should not be asking what these communications technologies will do to us – the right question is what we will do *with* them.

5

Build

For a moment, as I climb the steep steps of the Great Pyramid at Palenque in southern Mexico, I share some of the thrill of discovery that the Victorian explorer and archaeologist Alfred Maudslay must have felt when he came here for the first time. One hundred and thirty years ago, he was one of the first Europeans to reach the ancient Mayan city, at a time when the limestone temples and palaces were almost completely covered by impenetrable jungle. Even when I visit Palenque one humid December dawn, prying tendrils of foliage seem eager to reclaim this once mighty city. It is just one of the many cities of the Mayan empire that dominated most of Central America for 4,000 years, until its last city, Nojpetén in Guatemala, fell to the Spanish in 1697.

The Mayans' use of hydraulic cement was key to their versatile and durable architectural style. Although Palenque has been uninhabited and assaulted for more than a thousand years by tropical storms, high temperatures, earthquakes and the pervasive roots of tropical trees, most of its buildings remain structurally sound. The bright paint that decorated them has gone, but many of them are as imposing as they were during the city's magnificent heyday.

As it sets, cement undergoes an extraordinary transition from sloppy liquid to rock-like solid. When mixed with aggregate and water, it makes concrete, a metaphor for all that is robust and tangible. But concrete is not just prized for its resilient bulk – it can also be used to create soaring arches and graceful curves. Nearly two

FIGURE 5.1 I phone in Palenque. Might the Mayans have banned my mobile – or borrowed it? Mexico, 2016. Spot me.

millennia after it was built, the roof of the Pantheon in Rome remains the world's largest unreinforced concrete dome. The material has also been used to create the dense barrier of the Three Gorges Dam in China, as well as the half-mile-high Burj Khalifa tower in Dubai. It is also integral to many of the bridges that we travel over every day, and to the hidden sewers that take away our waste. The modern world is utterly reliant on the substance.

The ancient Mayans understood its importance, and learned how to build temporary blast furnaces, which could generate the extreme temperatures needed to transform crumbled limestone into cement[1] – a feat of chemical engineering that was not fully mastered in Europe until the middle of the nineteenth century.[2] The mortar used to build Palenque should last as long as the blocks of limestone that it bonds together.

The Mayans of the classic period (around AD 250 to AD 900) were keen observers of their world, and in particular the movement of celestial bodies,[3] and built their awareness of astronomical phenomena directly into their monuments – the stepped pyramid of the Temple of the Inscriptions at Palenque, for example, is perfectly aligned with the sun, and on the winter solstice the light of the setting sun descends directly into the heart of the temple. As Palenque's king Pakal the Great was carried into his tomb in AD 683, the architects of his memorial ensured that the sun would follow him down. This was precision engineering, executed on a grand scale.

I was sitting on top of Pakal's tomb when my mobile phone rang, wrenching me back into the modern world – a senior colleague needed to interrupt my trip to discuss an important deal. The sophisticated technology that allowed me to take the call served as a stark reminder that, despite the architectural, artistic and scientific sophistication of the Mayans, the choices they made about innovation resulted in their downfall.

Within a few generations of Pakal's death, Mayan civilisation began to unravel.[4] They were partly the victims of their own success; as their population grew, great swathes of jungle were beaten back and planted with maize, but the freshly exposed ground proved unstable – soil degradation and erosion, compounded by shifts in the climate that brought frequent droughts, made it increasingly difficult to grow enough food. In the eighteenth century, Thomas Malthus warned that food production would limit the expansion of all successful civilisations. In the case of the Mayans this was certainly true;[5] the archaeological record, carved friezes and the few Mayan codices that survived European looters tell a story of rapidly escalating violence and warfare, as different kingdoms fought over dwindling resources. The sharply hierarchical nature of Mayan society compounded these problems further as, rather than using their engineering prowess to improve the lot of individual citizens and to enhance the infrastructure of their cities and farmlands, the ruling elites continued to build grand monuments.

Throughout history, many powerful cultures have, like the Mayans, blossomed and then abruptly shrivelled. The magnificent relics of Palenque stand as testament to the fact that, when it becomes unshackled from human needs, progress can be undone more quickly than it is made. It is this sense of civilisation's transience that the great Persian poet and mathematician Omar Khayyam captured when he wrote, 'Think, in this batter'd Caravanserai, Whose Doorways are alternate Night and Day, How Sultan after Sultan with his pomp Abode his Hour or two and went his way.'[6]

INNOVATION CHALLENGES ORTHODOXY

As we plan how best to build the homes, communities and cities that can help our civilisations to flourish, we must avoid the complacency that proved the undoing of the Mayans. We also need to make sure that our buildings match their environment and are not vulnerable to the effects of our planet's changing climate. And critically, we must continue to build in ways that serve everyone in our societies, rather than just the most powerful. This was what the pre-eminent architect Richard Rogers, most famous for the Pompidou Centre in Paris, 3 World Trade Center in New York and the European Court of Human Rights building in Strasbourg, meant when he told me forcefully that 'All great buildings are modern in their time, so we must continue to innovate both technically and socially, in order to serve society.' I had visited Rogers at his London home. Wearing a chrome-yellow sweater, a blue shirt and neon orange braces, Rogers, now in his ninth decade, is pin-sharp and as iconoclastic as ever. He keenly understands the need for architecture to keep looking forward, rather than trying to mimic or enshrine that which came before, and he also understands that building in new ways can inspire fierce resistance.

The Lloyd's building in London is one of Rogers' best-known and most controversial designs. When it was unveiled in 1986, many people reacted as if an alien spaceship had landed in the heart of London's City. Now, many of the building's newer neighbours tower over it, but none of them have the same visual impact. Prominent metal ducts run up and down the outside to deliver air and water in and to take waste out, puncturing each of the fourteen storeys. Burnished steel staircases corkscrew up each corner, and illuminated glass elevators glide up and down in between them. Bright blue service cranes sit atop top a monumental set of grey concrete columns. All these different elements make up a building with an irregular foot-print, shaped by the medieval streets that surround it. Visually, it should not work, yet it does.

Rogers tells me that the Dean of St Paul's Cathedral once reassured him that he should not worry about the critics who were condemning his new building. When London's great cathedral was nearing com-pletion in the early 1700s, the architect Christopher Wren was in his eighties and in no mood to make any more compromises, having been working on the project for more than thirty years. So fed up was he with people criticising his plans that he asked his workmen to put up an eighteen-foot-high fence around the building site to

FIGURE 5.2 Built to stand, and to stand out: Richard Rogers' Lloyd's building, London.

conceal his revolutionary design as it took shape. St Paul's is now, of course, known around the world as one London's most venerable and striking landmarks.

Pioneering architects like Wren and Rogers are the quintessential innovators. Their new approaches to design and construction challenge society's preconceptions, and they have the power to change, often in profound ways, the way we live, work and interact. But, like the engineers behind all step-changing innovations, they understand that their creations will be accepted and last only if they fulfil real needs; and when buildings fail to do this, things can go awry.

Architectural history is marked by too many backward-looking and decadent vanity projects. Think of Mussolini's ugly attempts to recreate the grandeur of imperial Rome in a modernist style, or of Saddam Hussein's huge 'Mother of all Battles' mosque in Baghdad, an unrestrained celebration of a battle that he did not even win. This misguided approach to architecture is most clearly encapsulated by Albert Speer's grotesque plan to tear down Berlin and replace it with 'Germania', a regurgitation of classical forms on a monumental scale that would be a new global capital for Adolf Hitler's Third Reich.

However, St Paul's Cathedral and the Lloyd's building employ architecture in more constructive ways. For Wren, the human need served by St Paul's was clear; Britain's capital was reconstructing and repositioning itself after the Great Fire of London in 1666, which had destroyed two thirds of the City. Wren understood that Londoners needed a place to gather and an iconic symbol that would demonstrate their city's resurgence. The towering dome of St Paul's is an engineered masterpiece,[7] which remained the highest building in London until 1967. For its shape, Wren drew on the flamboyance of the Baroque, best exemplified by Michelangelo and Palladio's architectural masterpieces in Italy, but fused it with elements of the monumental Islamic style mastered by Mimar Sinan, the architect behind all the great buildings of the Ottoman Empire.[8] To support the huge edifice, Wren turned to more local and familiar structural techniques, using flying buttresses, which had been a hallmark of the previous era's medieval Gothic style. But rather than proudly displaying them, he concealed them behind screening walls, which in turn created the cathedral's grand and imposing facade. This radical fusion of architectural styles was a challenge to staid, Protestant Britain; many people wanted to look back and recreate the grand ecclesiastical style of the past, but Wren knew that London had to look forward.

With the Lloyd's building, Rogers helped both the insurance market and the City of London as a whole to enter a new future. Whereas Wren had gone to great lengths to hide his cathedral's structural elements, Rogers boldly displayed the building's vital organs on the outside. Neither architect was prepared to compromise in realising their vision. The latter's new style was mockingly nicknamed 'bowellism', but the approach served a function. Since its origins in Lloyd's Coffee House in late seventeenth-century London, Lloyd's has always been a space to discuss business and sign contracts; by removing the services from the interior of the building, Rogers created flexible, open spaces inside. The plunging central atrium, intersected by elevators, is an exhilarating void that is full of drama, but it is also surrounded by practical offices. The bottom two floors house a huge, bustling trading floor, where negotiating and underwriting take place. The Lloyd's building and St Paul's stand less than a mile apart, but they are separated by three centuries of architectural and engineering progress. They are very different structures, and serve very different needs, but each shows how our buildings shape the way we work, the way we meet, the way we live and the way we express power.

THE PROMISE AND THE PERIL OF THE CITY

As humankind began to put down roots and erect permanent buildings, everything changed. As farmsteads evolved into villages, towns and cities, it became obvious that these were places where knowledge could accumulate most readily, laws could protect and wealth could grow. The world's first true cities stood in Mesopotamia, the once verdant strip of land between the great Tigris and Euphrates rivers. The plough, the wheel, the sailing ship, arithmetic, cuneiform writing and metalworking were all invented here during the last 10,000 years.

Cities have throughout history proven to be the most stable features of our civilisations, often outlasting the nation states, kingdoms and empires that have ruled over them. They have, moreover, always been the primary sites of progress. The physicist Geoffrey West has created mathematical models that capture how the rates of innovation and wealth creation accelerate as cities grow.[9] Efficiencies of scale also mean that city dwellers use less energy per capita. As Richard Rogers neatly summarises, 'in terms of CO_2, it's much more efficient to take a lift than a car'. These advantages stem from the high density of people living in cities.

But improvements predicted by West's formulae do not accrue automatically. Density creates its own challenges; if people and goods cannot move freely within a city and if water, electricity and telecommunications networks cannot meet demand, progress will stall. As Paul Westbury, a director of the engineering and construction firm Laing O'Rourke, puts it, 'You also need to ensure that a city is liveable, that you can bring up a family there, that there's joy in the public spaces, that you have green lungs.' Or, as Rogers asserts, 'density should not be frightening, so long as it's well planned'.

We are fast becoming an urban species, so meeting the challenges associated with dense settlements is an urgent priority. 2008 marked the point when, for the first time, more than half of the world's population lived in cities[10] and by 2050, the urban proportion will increase to more than two-thirds of us.[11] Every year, new cities are built and the number of 'megacities' (informally defined as those that are home to more than 10 million people) continues to increase. Six hundred cities worldwide will be responsible for nearly two thirds of all global economic growth between 2010 and 2025,[12] and 440 of these are in developing countries.

NEW HEIGHTS AND NEW PROBLEMS
Many of the world's great cities, such as New York, Venice and Mumbai, have grown up on islands. Others, such as London, Rio de

Janeiro and Istanbul have been built in strategically important defensive or trading positions. Available land is always at a premium, so the natural solution is to build upwards. But, for many millennia, building more than a few storeys high was just not practical – the pyramidal temples of the Mayans, just like those of the Incas, the Aztecs, the Egyptians and the Buddhists and Hindus of southern and eastern Asia were exceptions. These buildings were much taller, but they were fabulously costly colossal monuments, designed only to inspire awe and signal prestige. That was also true of the medieval Gothic cathedrals in Western Europe, monumental structures designed to reflect the religious functions that they served.

The soaring vaulted ceilings, flying buttresses and imposing pointed arches of the best Gothic cathedrals are almost entirely self-supporting. Masons had to cut their stones with absolute precision because, unlike the Mayans and the ancient Romans and Greeks[13] before them, they did not have cement to bond their structures together. Even though medieval builders had lost the art of making true hydraulic mortars, their successful construction of these great churches undermines the common belief that technological progress stalled in the 'Dark Ages'. Etched into the floor of the crypt of Bourges Cathedral, for example, is a full-sized working template for the intricate stonework that was required to construct the windows of the nave, a humbling reminder that some of our predecessors possessed technical and aesthetic skills that cannot be matched today. In France alone, eighty cathedrals and thousands of smaller churches were built between 1050 and 1350.[14] Among them, the imposing forty-eight-metre-high arches of Beauvais Cathedral, the exquisitely balanced, gravity-defying vaulting of the Sainte-Chapelle in Paris and the peerless stained glass of Chartres Cathedral are undisputable masterworks of medieval art.

For centuries, these great medieval churches were outliers: extraordinary and incomparable architectural achievements that rose high above their neighbours. It was only at the end of the nineteenth century, when three technical advances coincided, that engineers could routinely construct bigger buildings. The first advance was the

FIGURE 5.3 Gravity-defying French architecture, still unmatched today: Bourges Cathedral (1324).

introduction of steel. Forging joists, girders, beams, trusses and other framing elements in dense, fire-proof steel with high tensile strength dramatically reduced the cost, difficulty and sheer amount of material required to erect taller buildings. The second advance we have already discussed: it was the availability of cheap, high-performance Portland cement. When cement is cast around steel supports, tall buildings can be built at an astonishing rate. However, for the upper storeys of a skyscraper to be useful, people have to be able to reach them. And so the third advance was the invention of the elevator, without which today's cities would look and feel very different.

Blocks and tackles, capstans, winches, windlasses and other forms of hoist were used to take builders and their materials to the top of the Gothic spires, and by the middle of the nineteenth century, steam-powered lifts were also commercially available. However, nobody trusted these devices to convey people safely up and down tall buildings, for if the load-bearing ropes were to fray and snap, people would plummet from huge heights. However, at an industrial exhibition in New York's Crystal Palace in 1854, Elisha Otis exhibited an innovation that would change the perception of the elevator and enable the first skyscrapers. Standing on a wooden platform that had been winched high above the crowd, he signalled that the hoist rope should be severed with an axe. As soon as the platform began to fall, a steel spring flicked out and locked with a vertical ratchet. Otis's safety brake brought him to a gentle halt, just a few inches below where he had started.

Modern lifts are much faster and more efficient, but the core of their design has changed little since 1854. Today multiple steel cables, each strong enough on its own to take the car's weight, are usually used instead of ropes. In some of the tallest buildings,

FIGURE 5.4 Braking news: Otis shows off his elevator safety brake at New York's Crystal Palace (1854).

including the Jeddah Tower in Saudi Arabia, which will for a brief period be the tallest in the world, carbon fibre-based 'UltraRope' will do the same job, and because it is lighter and stronger than steel, buildings can rise even higher. Instead of a mechanical spring, the safety brakes on most modern elevators are kept 'open' by powerful electromagnets, so if the lift loses power, the brakes immediately snap into place.

Structural steel framing, reinforced concrete and the elevator first came together to create a new kind of tall building in Chicago. Just as the Great Fire of London had allowed Wren to start anew in the seventeenth century, the fire that devastated Chicago in 1871 opened the city's eyes to new ideas and resilient building practices. The Home Insurance Building, commonly thought of as the world's first sky-scraper, was built in 1884. By today's standards, its ten-storey height is unremarkable, but at the time it was double the height of most com-mercial and residential buildings in the city. People could not comfort-ably climb up and down more than five storeys of stairs, still less when carrying deliveries and furniture. The use of steel meant that the building was not only fire-resistant, but also weighed less than a third of a brick-built structure of the same height. More importantly for the people working inside, the wall panels were not needed for struc-tural support, which allowed the architect, William Le Baron Jenney, to incorporate rows of large windows in his design. These 'Chicago School' buildings featured bright and airy workspaces, similar to those in today's skyscrapers. The virtues of this new approach to construc-tion were immediately obvious; it was a way to pack more businesses into Chicago's compact commercial heart. From the end of the nine-teenth century, the skylines of Chicago, New York and other major cities around the world stretched upwards.

High-rises gather in clusters to meet the need to pack ever more people into small areas of land; this can serve business and commerce well, and helps to keep city centres busy and vibrant, but they also change the experience of city living. When it comes to social housing, Rogers thinks that building high-rises is rarely the best answer: 'Once you're above eight stories, you've lost contact with the ground,' he says.

In June 2017 a fire broke out in Grenfell Tower, a twenty-four-storey 'Brutalist' housing block that had been built in West London in the early 1970s. The fire spread quickly and uncontrollably through the building, killing seventy-two people. Rogers makes the point that the disaster was ultimately caused by inequality, since 'we don't hear of wealthy towers burning – it's only the poor towers that burn'. The architect Neave Brown, best known for his low-rise social housing complexes in London and the Netherlands, agrees, in a recent interview he pointed to the fact that Grenfell Tower was hugely under-serviced and under-maintained. The building had just one staircase and two slow elevators. With better planning and maintenance, the fire could have been controlled and people could have been evacuated much more quickly. While disasters like the Grenfell fire are mercifully rare, Brown believes that high-rise social housing creates wider problems, by precipitating the breakdown of communities that live 'on very low incomes, with higher rents than they could afford to pay, borrowing to live in buildings that they didn't like'. Similar high-rise social housing schemes exist in rich and poor countries around the world, from the high-rise vertical slums of downtown Johannesburg to the faceless 'Khrushchyovkas' of post-war Russia. Very few are celebrated as desirable places to live, despite many attempts to make them more accessible and human; for example, the recently demolished Robin Hood Gardens estate in East London was built around so-called 'streets in the sky'. Brown condemns all these attempts – he says that they 'should never have been built, because they ostracise the underprivileged into special places for the poor, and therefore they become colonies of underprivileged people'.[15]

We have always needed shelter, and throughout history we have met the need for society's wealthiest citizens magnificently, while too often neglecting the poorest. The last thorough surveys estimated that some 100 million people worldwide are homeless, with more than a billion others lacking adequate housing.[16] This situation cannot be acceptable and it will only get worse – over the next three decades, the world's urban population will grow by an average of more than 1 million people every week. As Alejandro Aravena, the Chilean architect who won the Pritzker Architecture Prize in 2016, makes clear, if we do not provide everyone with appropriate housing 'it is not that people will stop coming to cities. They will come anyhow, but they will live in awful conditions.' This has implications for everyone. 'There is a ticking social time bomb being created in cities,' he says. Despite the scale of the challenge, Aravena has no doubt that the shift to urban life can ultimately be a good thing. 'If you measure every single indicator, from child mortality to access to education, jobs, health care and sanitation, all the numbers are better in cities than outside of them,' he says. Put simply, 'In order to reach people with public policies, it's more efficient if you have them concentrated in a city than sparsely distributed outside of them.'

Modern construction techniques allow construction companies to erect high-performance housing at a lower cost than ever before. The engineer Peter Head, former Director of Engineering at Arup, a planning and design consultancy, describes the dramatic improvement of modular houses that are built off-site, before being rapidly assembled where and when they are needed. 'It's conceivable now in a country like the UK, for example, that we could build 1 million affordable homes,' he says. They could, moreover, be built to such a high standard that they 'would cost nothing for people to live in, in terms of energy'. Aravena points out that the bulk of new homes globally will have to be built in the developing world, where cities are growing fastest, and explains that each house will have to be built for an average price of $10,000 per family.[17] To meet that target, builders will need to combine what Aravena calls 'high or even uber-tech' methods, such as 3D-printed plastics, with 'low or no-tech' methods, such as using mud brick, bamboo, rammed sand and reed roofing.

Living with real and present danger produces a more careful approach to standards, for everyone. I visited Mexico City in the autumn of 2017, shortly after an earthquake with a magnitude of 7.1 had rocked the city, destroying more than forty buildings and killing 220 people. However, just two weeks later the city appeared largely unscathed and was functioning normally. This was in stark contrast to the aftermath of the last major earthquake, which struck the city in 1985; that time, the earthquake was almost ten times stronger, with a magnitude of 8.0, and the damage was very much worse. Tens of thousands of buildings were destroyed, water, sewerage and electricity networks collapsed and over 10,000 people died. Mexico learned from that tragedy and devised prudent new building regulations, which applied the latest scientific understanding of seismic activity; as a result, all new buildings were much more resistant to earthquakes.[18]

BUILDING THE FUTURE

If concrete, steel and the elevator opened new possibilities for building in the late nineteenth century, the computer is the tool that has done most since the late twentieth century to change the ways in which we build. Tristram Carfrae, Deputy Chairman of Arup, describes the profound impact that computerised tools have had on engineering and architecture, but points out that their impact has been surprisingly gradual. At the end of the 1950s, Carfrae's predecessors pioneered the use of computers in structural engineering, using them to make the calculations that ensured that the spectacular 'gull's beak shell' roof of the Sydney Opera House was buildable. However, for a long time this was far from the norm; most calculations were still made by hand. As Carfrae explains, it was not until the 1980s that computer analysis became commonplace, but even then, it only led to incremental progress. 'At first, we got greater accuracy and probably reduced material consumption a bit,' he says. When, in the 1990s, the personal computer and the spreadsheet became ubiquitous, that 'still didn't really affect the outcome at all; it just changed the ease with which we could get to it'. There were a few exceptions to this, however. The spectacular glass-faced, fifty-seven-storey Torre Mayor in Mexico City, built at the turn of the millennium, is one of them. It easily survived 2017's powerful earthquake, because its engineers had used computer models to predict how seismic activity would interact with the tower's superstructure and the geological layers below. 'That kind of analysis didn't really start happening properly until the late 1990s,' adds Carfrae. The solution devised for the Torre Mayor involved ninety-six large, diamond-shaped dampers, which are on prominent display throughout the building. They act like the shock absorbers in a car, allowing the building to move and absorb the huge amounts of energy in an earthquake.

Since the turn of the twenty-first century, the impact of computer technology has taken the most significant of all leaps, as it has been transformed from a proving to an exploratory tool. Engineers have traditionally been trained to avoid making mistakes, at all costs. 'Our critical faculties are well attuned, because we've got to spot the flaw before we actually construct anything,' says Carfrae. That is why we can step onto a well-engineered bridge with the certainty that it will not collapse from under us. However, engineers now have the freedom to test new ideas in a virtual world, further liberating them from a purely functional approach. Carfrae describes the use of 'digital twins', discussed in Chapter 3, in architectural practice: 'I've always thought of it as three environments: a physical one; a virtual one that's connected to it, which receives information and issues instructions so as to configure the real one; and thirdly, a virtual test environment,' he explains. This last one is where engineers can experiment. Crucially, engineers can now make mistakes, so long as they make them virtually. The architect Norman Foster agrees whole-heartedly: 'There is no question that we can do things today that were not possible before being able to run computational models, antici-patory models, and to rapidly explore many variants. This is more than just an evolution – this is a new revolution and we're on the cusp of that now.' He looks forward to a new era of building innovation that is powered by digital tools assisting with the design and execu-tion of increasingly ambitious structures.

In 1943, while Winston Churchill was considering the best way to reconstruct the bombed ruins of the Houses of Parliament, he concluded that 'we shape our buildings and afterwards our buildings shape us'. In this new age of exploratory design and unprece-dented structural engineering capability, we are starting to see more buildings that self-consciously build upon Churchill's famous maxim and attempt to change the way people think and interact. Some are even designed specifically to foster a generation of innovative ideas.

From the outside, the Stata Center at MIT, designed by the American architect Frank Gehry, looks like the product of a collision between a jumble of irregular geometric forms. An assortment of different types of cladding, from red brick to brushed aluminium, accentuates the feeling of precarious imbalance. Inside, the spaces are equally contorted and irrational. Gehry hopes that these surprising architectural clashes will lead to innovative thinking, since the academics that work there will also collide in unexpected ways as a result. The Crick Institute in London is designed to achieve a similar objective, but its design takes a different approach. Opened in 2016, it has over a hundred laboratories, arranged around a huge central atrium. The building is designed to encourage the fluid movement of knowledge and skill between traditionally separate disciplines. Numerous bridges criss-cross the atrium and open into spaces for both formal and informal meeting. Apple's new corporate headquarters in Cupertino, California is built around a different concept still. It will house 12,000 employees in a serene and perfectly circular low-rise building that encloses an expansive botanical courtyard garden. Norman Foster, the architect behind this grand, gleaming loop of a building, emphasises his desire to create a 'bucolic' and inspiring space in the monotonous concrete sprawl of Silicon Valley.

Thomas Heatherwick has built on the idea of the workshop for the new London and California headquarters for Google. 'When you think of an office, something inside of you sort of slumps,' he says, 'so we put it to Google that we should try to build a workshop.' Like Foster's Apple building, Heatherwick's designs are arranged laterally rather than in towers. Private and public spaces blend with one another and with natural vegetation, creating a flexible and unconventional working environment. It seems likely that many more workplaces in the future will place a similar emphasis on adaptability and openness, as communications technologies improve and the boundaries between home and work life continue to blur. All these buildings have their own magnificence, and those that last will be our age's equivalent to the pyramids of Egypt and the great buildings

FIGURE 5.5 Green Apple: Take a bite. And a leaf out of Apple Park, Norman Foster's design for one of the most energy-efficient buildings in the world (2017).

of Palenque, though they will celebrate thought and life rather than ritual and death.

One important feature of many of these new buildings is their impeccable energy efficiency standards. Partly in response to a growing appreciation of the threat posed by anthropogenic climate change, they will all have a minimal impact on their natural environment. Foster's Apple building, for example, carries one of the world's largest rooftop solar power installations. He explains that it is also a 'breathing building'; shafts engineered into the concrete beneath the circular roof 'inhale' fresh air into the building passively, while warm, stale air is 'exhaled' elsewhere through chimney-like structures. It also has no need for powered air conditioning, heating or ventilation systems for at least nine months of each year. Foster hopes that these technologies and the environmental awareness that inspires them will catch on more widely. Rather than 'consuming at one end and spewing out waste at the other', he foresees a time when buildings and whole cities will efficiently convert waste into energy and when many buildings will generate more power than they consume.

A further development that Foster believes is poised to make cities more sustainable and resilient, and to help 'the whole economy move away from being linear and become circular' is urban agriculture. Peter Head is also optimistic about this idea. Highly efficient light-emitting diodes could provide artificial light twenty-four hours a day and be used within a city to grow high-quality fresh food, providing new sources of employment and simpler food delivery systems.

Buildings are no different from anything else that we engineer – they are becoming intelligent. When renovating my home in London, I tried to embed as much intelligence and automation within it as I could. There are no unnecessary switches, locks or visible consumer electronics. All this is meant to make life as easy and seamless as possible, but in reality these features sometimes simply do not work. Tony Fadell, co-founder of home automation firm Nest Labs, best known for their self-learning, sensor-driven thermostats, laughs when I tell him this. 'I look at this as the very earliest day of this world that we're creating,' he says. The real challenge of making intelligent buildings is not the individual components, but making them work together. Right now, Fadell explains, it may be no coincidence that many of these systems are designed by young bachelors who live alone – they tend to go awry when many people with different needs share a home or workplace. The ideal is that our buildings adapt and reconfigure themselves to the needs of their occupants at any given moment. This should all be seamless and automatic but, as Fadell emphasises, 'simple is hard'. He draws a parallel with the early decades of the twentieth century when electricity networks were first being rolled out, a time when everything from bulbs to light switches and grid connections were unreliable and unpredictable. Progress will catch up with the ambition, but, as Fadell concludes, 'all of these things take years to make robust'.

Norman Foster sketches a futuristic vision of the city, where buildings are sufficiently intelligent and self-sufficient to become 'independent of the physical grid of the city'. Not only will many of them be net producers of clean power, he also thinks they will be linked together by, for example, non-polluting electric vehicles and airborne delivery drones. Foster is visibly excited by the potential for the city to be transformed 'by quite an extraordinary combination of technologies into a highly liveable, desirable, clean place'. There is, however, a huge gap between constructing bespoke buildings that exhibit the very best standards of liveability, resilience and sustainability, and applying these ideas economically to a whole city.

ENGINEERING THE IDEAL CITY

Forty years ago, Shenzhen was a cluster of simple villages surrounded by green fields. Then, in May 1980, the Chinese leader Deng Xiaoping designated it a 'special economic zone', which opened it up to global trade. It was a bold experiment, but one that soon paid off; foreign investment flooded in, provoking rapid economic development and physical growth. Today, high-rise towers jostle for space and clusters of tall cranes herald the arrival of yet more buildings. Shenzhen itself is now home to more than 12 million people and has fused with the other fast-growing cities of the Pearl River Delta; the region has overtaken Tokyo as the world's largest continuous urban area.[19] If any single statistic highlights the extraordinary rate at which China is building, it is the fact that during the three years between 2011 and 2013, more concrete was poured in China than in the US during the entire twentieth century.[20] The factories of the Pearl River Delta produce a quarter of all Chinese exports, mostly electronics. The region has graduated from manufacturing mainly cheap household goods to being a leading centre for hardware and software innovation. Tencent and Huawei, for example, are both now colossal multinational corporations whose headquarters and research and development operations are based in Shenzhen. The province has a GDP equivalent to $1.2 trillion, which matches the output of several large countries, including Russia and Australia. Remarkably, the region's economy has been growing by at least 10 per cent each year for the last decade.[21]

Peter Head, who has contributed to many large-scale engineering and planning projects in China, has great admiration for the country's bold and meticulous planning, and lauds their systematic and highly integrated approach to managing energy, water, waste, transport and housing. It is all designed to improve the standard of life for citizens, as well as creating jobs and economic opportunities. Head notes that, 'No other country in the world is doing this,' and also observes that China is starting to take the lead when it comes to mitigating and minimising environmental damage. 'What they're doing is actually integrating human development with ecological development … China has now embedded the concept of the "ecological civilisation" firmly in its five-year planning cycle,' he says. On my recent visits to Shenzhen and Beijing, they have not struck me as particularly green or clean cities, but that is starting to change, and the amount of green space per capita is now going up rather than down.[22] Head describes the development of what is known as a 'sponge city', in which pavements and impermeable concrete barriers are replaced with porous structures, with all available space, including rooftops, central reservations and derelict areas, intensively planted to create natural and living water management systems. These judicious interventions will reduce pollution, mitigate the risk of flooding and reduce the impact of future climate change. They also deliver more pleasant and varied urban environments, which is important because, as Carfrae explains, 'there's increasing evidence that nature does things to our physiology and our well-being that nothing else can do'. For Head, the best way to build resilience is to engineer 'regenerative systems' rather than taking a 'protectionist approach, which is just trying to stop horrible stuff happening'. The idea of the sponge city illustrates this concept in a powerful way.

China's approach to urban development is based on the belief that rigorous design and testing will allow the creation of a great city from scratch. This has been tried many times throughout history, and it has sometimes worked. For example, in the eighth century, the Caliph al-Mansur built the original city of Baghdad; in the decades that followed, this perfectly round, walled city became a thriving centre for trade, learning and innovation. Often though, this approach fails. For example, at the end of the sixteenth century, the Venetian Republic built the fortified town of Palmanova; the shape of a nine-pointed star, it was enclosed by a high wall. It was designed to fend off attacks from the neighbouring Ottoman Empire, but also to be a self-contained city, which would house a self-sustaining and egalitarian population of merchants, farmers and artisans. The problem was that no one wanted to live there because it was deemed too isolated – eventually, the Republic resorted to housing pardoned criminals in their elegant new town.

Some modern-day 'concept towns' have had a similar fate. Songdo in South Korea was intended to be the ultimate 'smart city' – the authorities hoped that technology would make life there easy, safe and productive. Ubiquitous wireless sensors and cameras that are connected to centralised computer servers, monitor and opti- mise traffic flow and air quality, and identify suspicious behaviour.

FIGURE 5.6 Designed to be self-contained, but abandoned because of its isolation (Palmanova, Italy, 1583).

A system of pneumatic pipes sucks up garbage and removes it from the city. However, while everything works seamlessly, Songdo is under-populated and many of its buildings and public spaces are empty. Even when concept towns survive and are fully occupied, they sometimes fail to live up to expectations. Milton Keynes in the UK, Canberra in Australia and Brasilia in Brazil have never been exciting urban environments – they are bland and lack the heart and soul of their respective countries' more established cities. Other new cities are established primarily as experiments in new ways of designing and building and should be judged as such. Masdar City shimmers in the desert near Abu Dhabi in the United Arab Emirates. It was conceived as a zero-carbon, zero-waste paragon of sustainability, not, first and foremost, as a new home for large numbers of people.

Paul Westbury thinks there is an inherent risk that 'in striving for ultra-efficiency – you can accidentally iron out the wrinkles which make a place joyful'. He describes this as the 'new city conundrum'. Thomas Heatherwick believes that any approach which is too focused on utility risks neglecting the subtler design choices that mitigate some of the most negative sensations that built environments induce in us. 'I think we've lost a feeling for texture,' he explains. Different materials, whether ceramic brick, poured concrete, wood, steel or glass create 'those feelings in your stomach … that are sometimes more powerful than the shapes that look good in postcards'. This is a sentiment that the architect Amanda Levete clearly shares. When I visit her at her studio in North London, the receptionist asks me to leave my shoes by the front door; everyone in the building must pad over the thick red carpet in their socks – a simple act that reconnects them with the tactile sense and changes their experience of the space.

However, Levete is concerned that starting with a blank canvas can hinder the generation of the best ideas, regarding 'resistance as the fuel of architectural thinking'. That resistance comes from the local context, including accessibility, budget restrictions and the need to integrate buildings within a historical context. Historically, most cities have tended to grow gradually by absorbing existing settlements, which maintains their distinctive characters and communities. But when cities expand very rapidly or are built anew, that historical context disappears. 'When there's nothing there giving you clues as to why you might do something ... because there's no wrong way or right way, [there is a risk that] it becomes empty form-making,' explains Levete. She thinks this is the reason why many famous architects have done their least successful work in the new and growing cities of East Asia and the Middle East. In engineering terms, history is a constraint, but constraints can also stimulate the most creative solutions.

A BETTER WAY TO BUILD

'The best cities have a degree of chaos, utter non-pattern,' says Heatherwick. This is something that can make them dynamic and creative but it is, almost by definition, exceedingly difficult to plan and control. 'When designers set out major swathes [of cities], they tend to take away the vitality and real street life,' he adds. And too often, the people who will actually use new buildings are not involved in the decisions that inform their construction and use. Alejandro Aravena suggests a powerful way to return agency to citizens and nurture more diverse and interesting urban environments: 'What if your developer builds a base? It's got water, electricity and drainage. Then you [the inhabitant] do what you want.' His social housing projects put this concept into practice. He calculated that an average Chilean family needed eighty square metres of floor space to live well, while government subsidies were only sufficient to fund houses with forty square metres – his solution was to build 'half a house'. One half of these new dwellings is finished and ready for occupation, at the correct price; the other is built and left as an empty shell. The idea is that families will move in and save their money, only expanding into the unfinished half as and when they can they can afford to do so. This is a way to use good design to channel the energy and resources that may otherwise have gone into building slum dwellings into the creation of more stable and vibrant neighbourhoods. 'Instead of looking at slums as the problem, slums are *part* of the solution,' says Aravena. He thinks this model of 'porous urbanism', where cities are built and then left in an unfinished state, to be completed over time, is the only viable way to meet the world's looming housing crisis. 'When resources are scarce, do not build everything. Build what individuals cannot do individually,' he concludes, 'because individual action cannot, if not coordinated, guarantee a common good.'

When Aravena met Andrés Iacobelli, an engineer at Harvard University, they decided to do something to improve social housing in Chile. 'In my architect's mind, "do something" meant an exhibition or, in my wildest dreams, building a prototype,' says Aravena candidly. 'In his engineer's mind it meant, "Let's make a company that will compete in the market with the same rules but accept all the constraints and all the policies". And let's prove the market wrong.' So Aravena and Iacobelli founded a company with the ambition to show that social housing 'could perform as a tool to overcome poverty, not just as a shelter against the environment'. That is what happened when two families from their first 'half-house' scheme in Santiago, Chile, sold their homes. After twelve years, the families' initial $300 contribution had grown into an $80,000 asset. Aravena's scheme had housed 700 people per hectare, all in low-rise dwellings. That is nearly ten times greater than the city's average density, which means 'you can pay for land that is ten times more expensive,' says Aravena, 'and location is the number one reason for value gain.' So, rather than building isolated and under-serviced 'ghettoes' on the city's periphery, his company is able to build in desirable locations within reach of the city centre, which is where the opportunities for jobs, education, health care and recreation are concentrated. As Aravena points out, 'this is usually the reason people came to cities in the first place'. Engineering and design can come together with market forces to ensure that everyone gains from social housing. Aravena has shown the world a way to defuse the 'social ticking time bomb' that is being created by rapid urbanisation. Adopting this new model demands that builders, planners and investors all embrace a new mindset. Will it work? I ask Aravena. 'I'm reasonably sceptical and rigorously enthusiastic,' he replies with a broad smile.

Using architecture as an open platform is one powerful way to incorporate the aspirations of communities into the design process, but this process of embracing a wider diversity of ideas and expertise could be extended more broadly. The forms and functions of our buildings are too important for them to be decided upon by any one group of people. Even when a solution as radical as building 'half a house' is impossible, new developments, whether of domestic, commercial or public buildings, are most likely to succeed if their eventual inhabitants are consulted on every aspect of each scheme. Levete says that, at its best, architecture is 'the most open of all disciplines. It's cultural and it's artistic, but it's also about science; it's about politics; it's about society; it's about issues of national identity, personal

FIGURE 5.7 Buy one, get one half-empty: before and after shots of Aravena's 'half-house', Chile (2016).

identity and finding your place in the world.' That is the ideal, but it does not always work that way in practice. Norman Foster points out that 'architects are generally trained ... with the idea that the architect reigns almighty and can conceive a building and essentially hand it out to engineers to make it work'. This is a theory that has always struck Foster as utter folly: 'I have always encouraged a roundtable approach, where all the different disciplines come together in a non-hierarchical order ... For me, the ultimate [goal] is for them to fuse seamlessly together, in such a way that you can't unpick one from the other.' Westbury is in full agreement: 'The right answers come from the most diverse group of people that gather with different perspectives. You need urban planners, you need residents, you need business folk, you need academics, you need historians and you need engineers.'

Westbury laments the fact that this sort of integrated approach is only very rarely taken – in many parts of the world there are organisational structures that work against this way of thinking. 'If you look at construction,' he continues, 'the problem you've got is that the process is terribly fragmented.' New designs are conceived by architects and designers, who are often not well-versed in the delivery and engineering of buildings. 'They'd get one business to do steelwork, another to do concrete, another to dig holes, another to do systems and plant, another to do facades and cladding, another to do internal fit-outs and one to do glass.' For Westbury, this is a deeply inefficient and critically flawed process, because 'you only get the views of the people who are actually delivering the final product right at the very end'. Instead, he believes that their views need to be integrated much earlier, and he is optimistic that technology could be used to do this. He wants the industry to 'get away from the image of blokes in boots digging holes and to become a high-tech endeavour', suggesting that learning from the manufacturing sector and adopting modular construction approaches would be a good start. This would mean components and systems being made off-site in factories before being assembled on-site. Digital tools make it possible to fully specify their geometry and performance with unprecedented precision and accuracy; the certainty of delivery would replace the drift in design and cost so often seen in major projects.

Peter Head believes that progress can only be achieved with the better integration of skills and enterprises and thinks engineers are ideally equipped to help with this. In his view, the famous engineers of the Industrial Revolution were not technically excellent 'in terms of knowing where a bending moment was in a cast iron beam' but were rather 'holistic organisers of change'. Professional disciplines within engineering and beyond have become more focused and tightly defined and, though Head is concerned that 'there's a great "missing middle" of connectivity and collaborative decision-making', there are also powerful tools to catalyse integrated action. He describes a modelling tool that can allow people to test and demonstrate the social and economic impacts of new buildings; he hopes that this will become a 'seamless policy and decision-making tool' that will give investors the confidence to fund ambitious new projects. Ultimately, many decisions will still come down to human judgement, but with the involvement of such powerful simulation tools they will hopefully be better informed.

Richard Rogers is also clear that technology opens new possibilities for the way we can build, though he cautions that technology 'should always be your servant rather than your master'. 'All technologies can be used for good or for bad,' he continues, but 'you need to know about them and you need to use them in a positive way.' In the pursuit of shorter commutes and safer streets, poor design in the 'smart city' could encroach on privacy and remove some of the diversity and spontaneity of urban life. However, 'technology is not your enemy', Rogers urges, and 'the last thing you want to be is frightened of it'. Moreover, it need not be synonymous with complexity and great expense. After buying his traditional Georgian London townhouse thirty-five years ago, Rogers remodelled its interior completely, creating the huge, airy space, three storeys high, that we sit in to talk. Weak winter sunlight streams in through the room's huge sash windows. He reminds me that central heating is a technology that did not exist when the house was built and that without it we would be uncomfortably chilly on such a cold December day.

We tend to think of our buildings and cities as the most stable and permanent things that we have created but they are, in fact, all works in progress. Engineering will continue to change the ways in which we use spaces and the ways that cities work; buildings will become even more efficient, urban space will be used for agriculture and movement within cities will be streamlined. Rogers' transformation of his early nineteenth-century house is a case in point, and demonstrates that spaces are in constant flux. Shenzhen's transformation demonstrates the power of technology. Santiago's social housing scheme demonstrates how to avoid creating informal slums as more and more people migrate to cities, as well as how to avoid isolating them from the very services and opportunities that brought them to the city in the first place. Masdar City and Songdo may currently be under-populated, but there is much that we can learn from these urban experiments. The vine-covered ruins of Palenque and the smoke-blackened shell of Grenfell Tower, meanwhile, remind us that the only buildings that survive in the long term are those that work to address genuine human needs. All novelty builds on what came before; this is the way that progress takes shape and then gathers momentum. There is never a perfect solution and building is, after all, a process without any clear destination. The only way to proceed is to keep trying new things, keep listening to what people want and keep moving forward. As Rogers says, 'there's this tendency of always thinking that we are at the end of something. But it never ends.'

6

Energise

In 1962, the late US Senator John Glenn became the first American to orbit Earth. When he looked down, he saw that the dark side of our planet was decorated with a billion minuscule pricks of light; electric light is the only feature of our civilisation that is clearly visible from space. When seen from afar, all that distinguishes us from the rest of the universe is our ability to harness energy, transform it and use it to shape our world.

Energy is the central pillar of the human story and, throughout history, access to it has been essential for every single forward step. First we must consume calories, to feed brain and body. But everything that follows that primal imperative must also be fuelled, which is about far more than filling your automobile's gas tank. The shirts you wear, the medicines you take, the fruits you eat, the fashions you buy, the Web pages you browse and everything else you consume relies on the engineered systems that energise our world. Energy is human civilisation's greatest enabler, and without it everything would stop. We have always sought new sources of energy with which to amplify our efforts and increase our ability to bend the world to our will; the story of energy is about how it has released humans from the confines of their own muscle power, and ultimately, their own brainpower. This has happened very slowly and it is only in the last 1 per cent of our existence[1] that humans have uncovered a succession of increasingly potent energy sources and found ways to exploit them more efficiently, allowing us to build the vibrant societies that we live in today.

However, there is also a darker side to this story. Because it is so central to civilisation, the search for energy sources has been marked by commercial and political intrigue, corruption on a grand scale and global environmental despoliation. Oil-producing countries saw that wealth could be accumulated with little effort – in some, that encouraged laziness and corruption. Natural resources have been the cause of bloody battles in the two world wars – the fight over the coal in Alsace Lorraine in the first and the push to Baku for access to oil in the second. Oil has inspired the creation of cartels, such as OPEC, and monopolies, such as Standard Oil, and it gave birth to widespread laws that control the aggregation of corporate power. Humankind has not, for a long time, understood the full price of providing itself with energy – it has paid not only with its treasure but also with lives, broken moral codes and damage to our planet. Just as it was engineers who gave us access to the rich sources of energy that simultaneously liberated us and threatened our future, it is now engineers who must shoulder the responsibility of developing viable alternatives. And, if we are to write the next chapter of human progress, we need to gain access to yet more energy, while at the same time reducing our ecological footprint; that will mean dedicating more of our wealth and our ingenuity to keeping our world habitable in perpetuity.

My first encounter with the energy industry was as a small boy in the foothills of the Zagros Mountains in Southern Iran. My father was working for the Anglo-Iranian Oil Company, the earlier name for BP. We lived among the oil-producing wells in a small town called Masjid-i-Suleiman, where oil had first been discovered in the early 1900s. It was also the site of a Zoroastrian fire temple that had been built on top of a natural hydrocarbon seep, a clue to the vast oil reserves discovered in the area. The thing I most remember, apart from the searing heat, my warm Iranian friends, swimming in our swimming pool and parties, is the incessant roar of the vast flares burning off natural gas that was produced with the oil but had nowhere to go.

As I was about to go to university I decided that I should be self-sufficient from my parents – I had to earn some money. One of my sources of income was a university apprenticeship at BP. In the first long vacation I was sent to Llandarcy Oil Refinery for experience. I spent time in the technical laboratory, measuring the reduction in the viscosity of some extremely thick Libyan crude oil as other components were mixed with it, using something called a Ferranti-Shirley viscometer. Llandarcy has since been demolished and is now the site of Swansea University, but my interest in the energy industry is still very much alive.

After I graduated, I was persuaded by my father, who had by then retired, to join BP for a year's sabbatical from research to try the idea of working 'in a real job', as he put it. I found myself in an industry that, I later realised, was central to the human story. I entered it to solve problems rather than to create them – it is a decision that I have never once regretted or felt ashamed of, because I believe the industry can, and will, be a part of the solution to its own problems.

FIRE, EARTH AND WATER

The story of energy starts with our ancestors' discovery of fire. Occasional patches of scorched earth and charred animal bones appear in the archaeological record around 800,000 years ago. It took another half a million years for us to learn how to control fire reliably, but once we did, life improved, as the world became more controllable. Most fundamentally, cooking makes many foods edible and others more digestible, which gave our ancestors more calories and nutrition for less effort. Evolutionary scholars think the mastery of fire may have caused such a significant step forward because the extra calories released by cooking food drove the growth of our disproportionately large brains.[2] The foundations of civilisation were therefore built on food, with the development of agriculture occurring first in the Middle East around 9000 BC and soon thereafter, independently, on at least three different continents. Agriculture spurred the development of tools, which gave humans the capability to work more productively, and the use of domesticated animals further increased our efficiency. This resulted in the production of more food than one settlement could consume – trade and commerce were ways of creating advantage for those who controlled these surpluses.

Cities were built and hierarchies emerged. More and more labour was needed and much of this was provided by the peasants, while those with wealth, good health and influence rose to the top. The peasantry, on the other hand, had a poorer and less varied diet than their predecessors and living at high density in the new cities caused diseases to proliferate – as Pontus Skoglund, a geneticist at the Francis Crick Institute in London, notes, 'Researchers studying samples of ancient human DNA have detected a more radical evolution of the immune system [during the agricultural transition].' Stores of food and wealth were worth fighting over, and learning to make buildings and grow food may have made people more violent and bellicose.[3] The shift to agriculture demonstrates that innovation changes every part of life, including our biological make-up. Skoglund and his colleagues' research provides unambiguous evidence that humankind's dependence on engineered tools, and the ways of life that they enabled, has driven changes in the genetic instructions carried in our chromosomes and, therefore, the very foundations of our biology. Even today, he says, 'it's certainly the case that human populations are continuing to evolve'.[4] As we shape our tools, so they shape us.

The first attempts to make machines to help us do our work were based on wedges and levers – to split wood, to quarry slabs of stone and to move heavy objects. Then came wheels, pulleys, winches and gears. On their own, these inventions concentrated muscle power but did not expand it. Harnessing the kinetic energy of water and wind was a different matter – it was new energy. During the first century BC, the Roman engineer Vitruvius designed an efficient water wheel. The first windmills are thought to have appeared in Persia during the sixth century. The idea of using the energy of the sun to do our work had been around since antiquity, but making it work in practice was very challenging. According to Greek legend, as Roman ships advanced on Syracuse in Sicily, Archimedes ordered soldiers to use their polished shields to focus the sun's rays onto their sails and ignite them. Leonardo da Vinci imagined a more peaceful use for solar power – he designed, though was never able to build, huge arrays of silver-coated glass mirrors that could boil water and melt metal ores. Solar power would have to wait. However, the improvement of water and wind power was more straightforward, and these were widely used by late medieval times to grind flour and saw timber. In 1540, Vannoccio Biringuccio, the pre-eminent metallurgist of the Italian Renaissance, highlighted the importance of water power to the industry of the time: 'Of all the inconveniences, shortage of water is the most to be avoided ... because the lifting power of a water wheel is much stronger and more certain than that of a hundred men.'[5]

Powerful as they were, these energy sources had severe drawbacks. Wind is fickle and unreliable, and water power is limited in its geographic reach. Up until the beginning of the nineteenth century, the energy from human muscle, augmented by domestic animals, the windmill, the water wheel and heat from burning wood, charcoal,

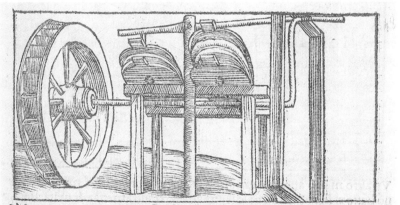

L'altro modo fie facendo fimile alle fopradette vna ruotta & in capo del biligo fia vn fimile affe, & fopra alli mantaci fia vna trauerfa biliga- tache a vna tefta habbi vn contrapefo, & da l'altra fia el manicho che e:

FIGURE 6.1 Going round and round: generating kinetic energy through a water wheel. From Biringuccio's *De la Pirotechnia* (1540).

animal dung and any other available biomass, were all that societies had to help them do their work.

COAL AND MOTION

Coal changed everything. It had been used on a small scale since antiquity,[6] but to Marco Polo it was a new wonder. Visiting China in the thirteenth century, he described 'a kind of black stone existing in beds in the mountains, which they dig out and burn like firewood. It is true that they have plenty of wood also,' he continued, 'but they do not burn it, because those stones burn better and cost less.'[7] The power of coal was plain to see, but it remained a marginal energy source because most of it was buried deep under the ground. It had to be mined, which is where the steam engine first came to be of use; inefficiently powered by coal, they were used to pump out the water that seeped into coal mines. Increasing yields of coal meant that iron foundries, brick kilns and other energy-intensive enterprises scaled up their operations, further increasing the demand, and a positive feedback loop began. Improvements to the steam engine meant that more coal could be extracted from mines. More coal meant more industry, more wealth, a drive to create new businesses and the need for even more energy. And coal was the catalyst for it all, since the amount equivalent to the weight of a man can produce the same energy as the same man could if he worked for a hundred days.

There was a widespread fear of coal-powered machines and what changes they might cause. Many believed that they would replace human labour and create mass unemployment, but they instead created jobs by the thousand. In Britain, people came from the countryside to work in the mills and factories of the towns. There was plenty of work to be had, but the hours were long and gruelling and living conditions often cramped and squalid – Arnold Toynbee, the historian who first coined the phrase 'Industrial Revolution', condemned the early nineteenth century as 'a period as disastrous and as terrible as any through which a nation ever passed'.[8] But there was also a brighter side – wages rose steadily in the second half of the century, as did average height and life expectancy, all measures of a healthy life.[9] Though contemporary accounts by workers[10] describe the cruelty inflicted on children whose lives were often cut short by their work in textile mills and mines, few adults yearned to return to the rural existence of meagre wages, servitude, fuel shortages and hunger. Despite the long shifts, noise and filth in the factory, 'industrialisation had a remarkable power to put food on the table'.[11] Governments recognised, for the first time, that the new industries should be regulated to ensure that technological progress translated into societal progress. Wages rose steadily and the education and skills needed to join the swelling middle classes became more accessible. The world was growing, and coal was at the root of it all.

As coal became an abundant and accessible source of power, Britain became the world's richest and most influential country. The economist William Stanley Jevons was the first to acknowledge its critical importance, writing in 1865 that 'coal in truth stands not beside but entirely above all other commodities'. However, he also warned that Britain's downfall would be inevitable if it continued to rely entirely on coal,[12] because reserves of it were inarguably finite and, since demand was increasing at an exponential rate, it was only a matter of time before mines were depleted and coal became prohibitively expensive. His ideas coincided with those of Thomas Malthus, the philosopher who believed that, because population growth would always exceed the growth in food production, civilisation would inevitably revert to a life of subsistence and toil. Both Malthus and Jevons failed to understand that engineering, combined with new sources of energy, would provide ways for the production of goods and natural resources to keep pace with, or even exceed, the growing demand. Jevons was even sceptical of innovation, scorning the 'notion very prevalent that in the continuous progress of science, some substitute for coal will be found – some source of motive power as much surpassing steam as steam surpasses animal labour'. He was, of course, proved wrong by the discovery of oil.

OIL: A MORE POTENT SOURCE OF ENERGY

No other combustible energy source contains the same power, gram for gram, as crude oil. However, by itself it is of little use – a mix of

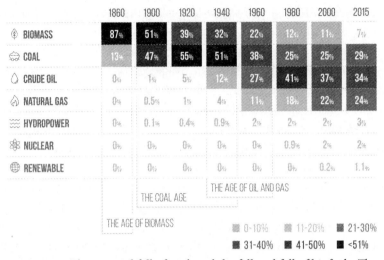

	1860	1900	1920	1940	1960	1980	2000	2015
BIOMASS	87%	51%	39%	32%	22%	12%	11%	7%
COAL	13%	47%	55%	51%	38%	25%	25%	29%
CRUDE OIL	0%	1%	5%	12%	27%	41%	37%	34%
NATURAL GAS	0%	0.5%	1%	4%	11%	18%	22%	24%
HYDROPOWER	0%	0.1%	0.4%	0.9%	2%	2%	2%	3%
NUCLEAR	0%	0%	0%	0%	0%	0.9%	2%	2%
RENEWABLE	0%	0%	0%	0%	0%	0%	0.2%	1.1%

THE AGE OF OIL AND GAS

THE COAL AGE

THE AGE OF BIOMASS

- 0-10%   - 11-20%   - 21-30%
- 31-40%   - 41-50%   - <51%

FIGURE 6.2 The rise and fall of coal, and the fall and fall of biofuels. The number of energy sources on the rise since the beginning of the twentieth century.

many hydrocarbons, it needs to be 'refined', which originally meant distilled into different fractions. One product of distillation, kerosene, was a driver of the first US oil boom. During the first half of the nineteenth century, whale oil was used widely for lighting, but by the middle of the century, whale populations were dwindling and the price of lamp oil was rising.[13] Kerosene, derived from crude oil, filled the void; it burned more cleanly and brightly, but without the fishy smell of smouldering blubber, and it was cheaper.

The oil boom was ignited in 1859, when Edwin Drake drilled the world's first commercial oil well in Pennsylvania. He stimulated a surge of investment in oil exploration and production, which led to the discovery of large deposits of crude oil below ground level. It became the most important of all energy sources, since it gave us increased mobility and freedom. During the twentieth century, the ever-increasing global population wanted more and better ships, automobiles, airplanes and petrochemicals; thanks to oil, all of this was made possible, and at very low cost.

At first, fulfilling the growing demand was a relatively simple matter of drilling wells into likely accumulations of oil. The structures that contained oil could often be seen mirrored on the surface of the Earth. I remember this vividly when, as a child, I flew over the Zagros Mountains, the range which contained the oil reservoirs that established BP, to my new home in Iran.

Just as the doubts crept in over coal reserves, people soon began to worry that oil reserves would run dry. In 1914, the US Bureau of Mines reported that the country's oil reserves were just 6 million barrels, around 6 per cent of the world's daily consumption of oil today, and would soon be used up.[14] In 1956, the American geologist M. King Hubbert coined the term 'peak oil'. Revisiting Jevons' fears, he made a firm and widely publicised prediction that oil production would peak around the year 2000, before declining rapidly. His forecasts proved to be spectacularly wrong – apart from a brief blip after the 2008 global financial crisis, global oil production has risen

FIGURE 6.3 Blowing out without blowing up: the first oil well in the Middle East, Masjid-i-Sulaiman in Iran (1908).

steadily since the turn of the millennium. When I joined BP in 1966, the world had less than 400 billion barrels of proved oil reserves; at the end of 2016, that figure had quadrupled to over 1,700 billion barrels, despite the production of at least 1,200 billion barrels during the intervening fifty years.[15] The production of oil has continued to meet and has often exceeded the demand for it because new engineering solutions have allowed more oil to be found and produced, even in intimidating geographies such as the High Arctic or from very deep water such as the Gulf of Mexico.

To produce oil safely from these areas, without spilling it, is an extraordinary engineering feat. I started my career in Alaska, where I worked some 250 miles above the Arctic Circle. Here oil had to be extracted from 10,000 feet below the surface, and beneath 1,600 feet of permafrost. At the other extreme, during my time as CEO, BP developed an oilfield called Thunder Horse in the deep water of the Gulf of Mexico. Oil and gas are produced there from wells on the seabed, which are connected to a platform floating in 6,000 feet of water – it is a monumental structure that can handle the production of 250,000 barrels of oil per day. Ambitious engineering projects can also fail with disastrous consequences for human life and the natural environment. From the Piper Alpha disaster in the North Sea in 1988 to the explosion at the BP Macondo oilfield in the US in 2010, these

FIGURE 6.4 Tackling man-made disasters isn't just about putting out the fire. Here, after the explosions on BP Macondo oil platform, April 2010).

incidents remind us that, when it comes to providing energy, companies have a heavy responsibility to navigate between the possible and the desirable, because the consequences of failure can be so great.

NATURAL GAS: A GREATER FUTURE

Oil was changing the course of history. There was, however, little appreciation of its damaging contribution to local pollution and global climate change, the hidden and unintended consequence of progress. As the oil industry expanded, engineers realised that natural gas, which had often been regarded as a useless by-product of oil extraction, was a potent and convenient energy source in its own right. Gas had been used before, in the form of 'coal gas', which was especially effective for street lighting, but natural gas was easier and cheaper to produce than gas from coal and, as more pipelines were built, it fed factories and power plants and provided heat for homes and businesses.

Even more importantly, natural gas helped to sustain the dramatic population growth during the twentieth century. In 1909 the German chemist Fritz Haber made hydrogen from natural gas and reacted it with nitrogen to make ammonia, the crucial ingredient of artificial plant fertilisers. Then an engineer called Carl Bosch devised a way to make the process work on an industrial scale. The Haber-Bosch process brought abundant and cheap chemical fertilisers to the world, quadrupling the productivity of agricultural land.[16] Haber was also one of many deeply tarnished inventors – when the First World War broke out, he enthusiastically joined the German war effort and developed poisonous gases that killed and maimed thousands of Allied soldiers. Meanwhile, the ammonia made by his process was used as a precursor to nitrate, a crucial ingredient in Germany's bombs. According to Churchill, 'but for the invention of Professor Haber, the Germans could not have continued the War after their original stack of nitrates was exhausted'.[17]

Natural gas has clear advantages as a fuel and feedstock – it burns much more cleanly than coal or oil, and it is also very abundant. Recently, two remarkable engineering innovations have coincided, to ensure that natural gas will be an abundant fuel long into the future. First was the creation of cost-effective technology that can liquefy natural gas by chilling it to around minus 160 degrees Celsius – by reducing the volume of the gas by 600 times, it has become possible to transport it in ships from countries with a surplus to those with a shortage. Second was the introduction of hydraulic fracturing, often referred to as 'fracking', combined with horizontal wells drilled into almost impermeable shale layers. High-pressure water creates fractures in the shale, which are then propped open with sand or ceramic balls, allowing gas to flow out. Horizontal wells, meanwhile, open up a much larger area of the rock formation than a vertically drilled well. When the two innovations are combined, a great deal more gas can be produced from each well. This has given the US close-to-infinite gas reserves and has also made the country one of the world's top two crude oil producers, transforming the global oil market.

As with many other new engineering advances, fracking has inspired fears as well as creating benefits. The movement of large amounts of heavy equipment, the use of large volumes of water with added chemicals[18] and the increased occurrence of earthquakes have left many communities suspicious of the comforting words issued by oil and gas companies. The chemicals used are described as harmless but their exact nature has very often been kept secret. There are fears that fracking will break into potable water aquifers and pump them full of these chemicals. If fracking takes place near natural geological fault lines, earthquakes can be induced; most are small, roughly equivalent to the rumble of a bus passing by, but just occasionally they are much larger and cause damage to homes. As a result, some communities have taken a precautionary approach and prohibited fracking. In 2016, there were 1 million producing wells in the US, two thirds of which were horizontally drilled and hydraulically fractured.[19] The evidence from this activity seems to indicate that fracking, when performed carefully, is safe.[20] However, rational explanations and statistics are not enough to resolve such an emotional and polarising issue, which will likely be limited to a few very

FRACKING AND HORIZONTAL DRILLING

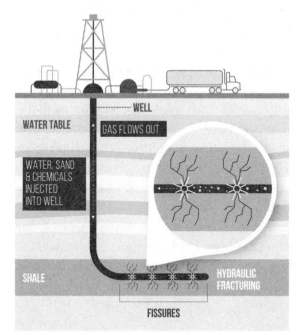

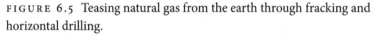

FIGURE 6.5 Teasing natural gas from the earth through fracking and horizontal drilling.

sparsely populated areas – when industrial activity is out of sight, it is usually out of mind.

LIGHTING UP THE WORLD

In a small display to one side of the Energy Hall of London's Science Museum is a piece of engineering that delivered energy in a new and incomparably flexible form. The Irish-born engineer Charles Parsons was unimpressed with the efficiency of even the finest steam engines of the late nineteenth century and believed that he could do away altogether with the pounding pistons and crankshafts. He replaced them with the invention that is in this display case: the steam turbine.

Parsons' turbine-powered boat arrived uninvited at a naval fleet review organised to celebrate Queen Victoria's diamond jubilee and promptly stole the show, shooting past the flotilla at 34.5 knots

(64 kilometres per hour). To the delight of many observers and the frustration of the Royal Navy, SS *Turbinia* proved itself much faster than any other vessel afloat that day.[21] Not for the first time, a single-minded entrepreneur had trumped organisational inertia, and Parsons' turbine has had a profound impact on the world ever since. The high speed and smooth, steady rotation of the turbine made it ideally suited to electricity generation – when paired with the dynamo, which converts motion into electric current,[22] the steam turbine began to change the world. Even today, electric power plants that are fuelled by coal, natural gas or nuclear fission depend on steam – at their core, they are all descended from Parsons' engineering breakthrough.

Electricity offers us endless ways to use energy. It can be transmitted over long distances, and in a high voltage direct current form, which can be done with little loss of efficiency. We can turn it on

FIGURE 6.6 The SS *Turbinia* (1894). Turbine power made the steam engine run out of steam.

and off, or store it in batteries for later consumption. And crucially, electricity can be readily translated into light, heat, refrigeration or movement. The electrification of our homes, cities, schools and industries was, I believe, the most profound technological transition of all time, and one of the greatest advances in civilisation.

Starting in the UK and the US at the end of the nineteenth century, the electricity revolution spread across the world quickly, but inconsistently. Sam Goldman, the ambitious young founder of a company that sells solar-powered lamps in developing countries, is proud of the story of one of his customers in central Kenya. Every day, Agnes Wanjiru gets up before dawn and travels to a small town, where she earns a living by selling potato chips on a market stall. Early in 2015, frustrated by the dim light, choking fumes and expensive fuel for her kerosene lantern, she borrowed enough money to buy a small, solar-powered LED lamp – as a result, her income nearly doubled, because she could fry more potatoes and her customers could better see what they were buying.

Agnes was one of the billion people in the world with little or no access to electricity; another billion can only access a grid that is so patchy and unreliable that they are essentially reliant on burning kerosene, wood and biological waste. Cheap, reliable solar powered lamps are bringing electric light to energy-poor people across the globe. Cambridge University Professor Richard Friend tells of the recent technological progress that has made this possible: 'There's been a factor-of-ten improvement in every single element of these new lamps; a tenfold reduction in cost of the solar cell; a ten times improvement in the efficiency of the LED, compared to the tungsten light bulb; and the key quality of the lithium ion battery is that it can be fully recharged and discharged ten times more often than one could do with lead acid batteries.'

As Agnes's story demonstrates, lighting can change lives profoundly. By extending the useful length of the day, electric light creates time for reading, education and growing businesses, which can immediately enlarge the prospects of millions of people in the poorer parts of the world. It is hard to overstate the impact of electric light, and even harder to comprehend why this 150-year-old

FIGURE 6.7 An LED-led revolution brightens up the world: children in rural Bihar, India, delighting in a solar-powered LED lantern.

innovation remains unavailable to so many people. Just as electrification represents one of the greatest advances in civilisation, continued energy poverty remains one of civilisation's greatest failures. Agnes's story provides a glimmer of hope that the current energy transition will finish the task that was started so long ago.

CLIMATE CHANGE AND THE GREAT TRANSITION

The economic historian Robert Gordon makes the point that the 'special century' between 1870 and 1970, when global economic growth took off, was only possible because of a unique clustering of 'great inventions', which included electricity generation and the internal combustion engine.[23] During this period, prosperity and living standards doubled for each subsequent generation of US citizens. Innovation gathered pace, education became widespread, health care improved and civil rights advanced. And the engineering triumphs that Gordon hails were both, of course, powered by fossil fuels – it was coal, oil and natural gas that finally proved Jevons and Malthus wrong and propelled humanity from subsistence to opportunity and growth.

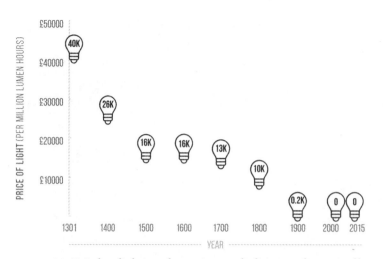

FIGURE 6.8 Priceless lighting: dramatic cost decline is a glimmer of hope for people in developing countries.

But the story of fossil fuels does not end there. Even as they were driving rapid progress, they were also forcing an invisible change in our planet's atmosphere. As we burn fossil fuels, the carbon contained in them is released as carbon dioxide; since the Industrial Revolution, the amount in the atmosphere has increased by 50 per cent, from 280 to over 400 parts per million, making it the most abundant of the so-called 'greenhouse gases'. Another is methane, most of which is emitted during the production and transportation of coal, natural gas and oil – although it is much less abundant than carbon dioxide, it is a more potent greenhouse gas in the short term. As these gases build up, they trap heat in our atmosphere, increasing the average temperature of our planet. This change is progressively altering the climate, and has serious consequences for our ability to inhabit our planet.

In Washington, DC I meet James Garvin, Chief Scientist at NASA's Goddard Space Flight Center. On a huge screen he shows me a series of time-lapse images of the North Pole, as seen from space. Each summer, blue fingers of seawater invade the delicate mass of pure white ice, and each winter the ice grows back, but over the three decades that NASA has been watching, the ice cap is steadily losing

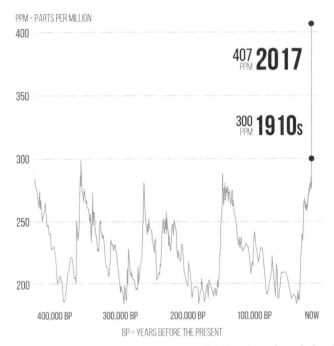

400,000 YEARS OF CARBON DIOXIDE

PPM = PARTS PER MILLION

407 PPM **2017**

300 PPM **1910s**

400,000 BP   300,000 BP   200,000 BP   100,000 BP   NOW

BP = YEARS BEFORE THE PRESENT

FIGURE 6.9 Concentration without precedent: carbon dioxide levels since the Industrial Revolution have gone off the charts.

ground. 'Back in the late eighties, we used to see big, red-covered ice floes around Greenland, where polar bears had slaughtered seals,' says Garvin. 'They've all gone. We are in a state of profound change.'

Naturally, most of those who produced and sold the fuels whose combustion emitted the most carbon dioxide denied the link between their actions and the changing climate, long after the evidence clearly indicated otherwise. It was a little like those 1960s American advertisements that showed 'physicians' in white coats promoting one brand of cigarette over another on the basis of spurious health benefits – vested interests have a powerful way of blinkering people to inconvenient truths.

In May 1997 I was the first leader of a major oil company to break ranks with my peers and acknowledge the link between the burning of fossil fuels and climate change. Speaking in the open-air Frost Amphitheater at Stanford University on a hot afternoon, I made the case that we should start the long and arduous process of weaning ourselves off our reliance on fossil fuels. Two decades on, that action is even more urgent. Garvin's graphic illustration of retreating ice is the most visible manifestation of global warming, but it is not the only one. When oceans get warmer, they absorb more carbon dioxide, making the water more acidic. As a result, many of the world's coral reefs are bleaching and dying.[24] Across the globe, people are reporting seasonal changes to weather patterns, plant growth and animal behaviour that are outside their previous experience.[25] An overwhelming majority of climate scientists agree that climate change is happening and that the chief anthropogenic driver is the extraction and burning of fossil fuels. The precise consequences of failing to curb greenhouse gas emissions are hard to predict with complete confidence. At the Weizmann Institute of Science in Rehovot, Israel, I meet Dan Yakir, Professor of the Earth and Planetary Science Department. As we walk around the sun-dappled campus, he takes me through the complexity of our climate: 'If we plant a dark forest on white sand dunes it is considered a big success, since it absorbs some carbon. But the focus on carbon can be sometimes misleading because we're not interested in carbon *per se*, but because it influences the energy budget of the Earth's surface.' The radiation energy that a new forest absorbs would have otherwise been reflected back to space by the white dunes, which is known as the albedo effect. Forests create clouds by drawing water from the earth and evaporating it but, as Yakir explains, 'clouds are also tricky as climate is sensitive to the type of clouds that you create. High clouds actually enhance warming and low clouds enhance the albedo effect.' Nothing is as straightforward as it seems.

How human populations will react to the changing physical geography that will result from climate change is even harder to forecast. Most models suggest that sea levels will rise and extreme weather events will become more frequent and severe, while serious droughts and crop failures will be increasingly common. If these forecasts are correct, climate change may well cause mass migration from the worst affected areas, which would lead to widespread political instability and huge humanitarian and economic costs. 'Ultimately, I think the rich will be able to adapt [to climate change], but the poor will have difficulties,' concludes Yakir. Discussion of the consequences of climate change is now very common. At the 2018 Venice Architectural Biennale, for example, one of the most popular exhibits was a simulation of the consequences of climate change on sea levels and inhabitable land. Discussion is, of course, important but we must act to secure the energy we need to maintain growth and opportunity for all, while also reducing greenhouse gas emissions. I am confident that this can be done.

On a chilly spring morning in Boston, I visit MIT's Professor Ernie Moniz, US President Barack Obama's former Secretary of Energy. Moniz wants to tackle the entangled problems of climate change and the maintenance of adequate low-cost energy supplies. Just like James Lovelock, the distinguished environmentalist, he is refreshingly unsentimental and makes the point that the key objective is to save human lives, rather than planet Earth. Our planet will adapt and persist, argues Lovelock, but there is a real risk that humankind will not if it does not change its ways quickly enough. Moniz starts by discussing natural gas. When burned, gas releases about 50 per cent less carbon dioxide per unit and much fewer toxic air pollutants than coal. Since 2005, US carbon dioxide emissions have dropped by 13 per cent.[26] Much of this reduction comes from the replacement of coal-fired power plants with ones that burn the new, cheaper reserves of gas, produced by fracking and horizontal drilling. The substitution of coal by gas in the US has arguably been the most important direct contributor to decarbonisation so far this century – and it came about not as a result of policy or government intervention, but of technology and market forces.

'There's no question that for now natural gas is part of the solution in the US,' says Moniz, 'but eventually it becomes part of the problem.' He thinks of it as a 'bridging fuel', an energy source that will ease the transition to a low or zero-carbon economy. Using more gas, instead of coal or oil, is a direct and proven way to achieve immediate reductions in greenhouse gas emissions. Moniz remains optimistic that better ways to capture and use the carbon dioxide emitted from natural gas will continue to be developed; if that happens, the world has immense reserves of gas that could provide clean power for generations to come.

THE DREAM OF NUCLEAR: PAST AND PRESENT

Many have long sought a universal fix for the world's energy needs, which is what the physicist Arthur Eddington believed he had discovered when he identified the source of the sun's power back in 1920.[27] Eddington correctly inferred that massive releases of energy occur when hydrogen atoms crash together and fuse to create atoms of the heavier element helium.[28] He wished that we could create the same source of energy on Earth; a gram of hydrogen fuel could theoretically generate up to 10 million times more energy than a gram of coal. The fuel is abundant and cheap[29] and its main by-product, helium, is harmless.

Steve Cowley greets me at the Joint European Torus (JET) fusion reactor at the Culham Centre for Fusion Energy near Oxford in the UK,[30] an establishment he previously directed, and gets straight to the point. 'Critics of fusion say that you can't put the sun in a bottle. But that's the bit we've done.' He leads me into a huge hall, and explains that the remaining challenges are about engineering rather than physics. Behind a dense mesh of pipes, girders, fat cables, cranes and scaffolds, I can see the outline of a large, doughnut-shaped structure, about the height of a two-storey house. Inside it is a swirling cloud of ionised plasma,[31] which reaches a temperature greater than 300 million degrees Celsius, twenty times hotter than the core of the Sun. So far, the most powerful fusion reaction it has achieved consumed 50 per cent more energy than it produced.[32] The reaction fizzled out within seconds because of turbulence that snuffs out the reaction;[33] no one has yet figured out how to prevent this. JET will soon close and work will shift to an even larger and more complicated reactor that is currently being built in the South of France. The goal of the ITER (International Thermonuclear Experimental Reactor) is to produce ten times more fusion power than it consumes, but there can be no guarantee that it will succeed. Cowley explains that the reactor will be surrounded by 2,800 tonnes of the biggest superconducting magnets ever built, which are so large that they cannot be fully tested until they are built into the reactor. 'If something doesn't work, we'll have to take whole thing apart,' explains Cowley. There is no room for trial and error without huge cost. And as the scale, complexity and cost of any project escalates, engineers no longer innovate because they can take fewer and fewer risks as they search for solutions. As Cowley concludes, 'You can't be Thomas Edison if every step is going to cost you a billion euros.' As I say goodbye, I ask him when he thinks fusion will be a viable commercial proposition. 'I think it will be making a meaningful contribution by the end of the twenty-first century,' he answers. I leave doubting that fusion is the timely 'magic bullet' fix for our future energy needs; we will need dramatic progress in other areas to forestall the effects of climate change.

Although engineers have yet to turn nuclear fusion into a viable power source, nuclear fission, the release of nuclear energy by breaking atomic nuclei, was mastered two generations ago. The world witnessed the first dramatic demonstration of nuclear power when atomic bombs were dropped on Hiroshima and Nagasaki; engineers realised that they could better control the nuclear fission reaction, to make the heat needed to generate an abundance of electricity. This is now a proven technology, which makes very low net contributions to greenhouse gas emissions. In principle, fission could make a great contribution to the low-carbon energy systems of the future. However, in March 2011 the world was reminded why many people are wary of this energy source, when a magnitude nine earthquake sent a powerful tsunami barrelling into Japan's north-east coast, heavily damaging the Fukushima Daiichi Nuclear Power Plant and forcing an evacuation of the surrounding area. The Japanese government estimate that the clean-up will take at least three decades and cost almost $200 billion, and in the aftermath of the incident all of

Japan's nuclear reactors were switched off. The disaster highlighted the risks associated with nuclear power generation, but viewed as a whole, the safety record of the nuclear fission industry is actually very good, and almost certainly improving. Over the last six decades, thirty-three countries have installed fission generators, which have been operational for a total of 17,000 years. In this time, there have only been three major nuclear incidents: Fukushima Daiichi; Three Mile Island in the US in 1979; and Chernobyl in Ukraine in 1989. Of these, only the Chernobyl incident resulted in the loss of life. Ernie Moniz shares my hope that the nuclear power industry can re-kindle public confidence and make a bigger contribution to a low-carbon future. He is optimistic about a new generation of small, modular fission reactors, with entirely passive cooling mechanisms. If reactors were built to a standardised design and produced at sufficient scale, their costs would drop dramatically. Moniz points to several American companies that are making small reactors that could travel from a factory to their destination on the back of a truck. This is a very different approach from the huge, complicated, one-off designs being built elsewhere. Richard Friend explains that this cumbersome method 'defies the important law of engineering, which is that costs fall at 2 per cent per annum compound, because we get smarter at making the same thing'.

The work on these small reactors sounds promising, but none of them yet has regulatory approval, and the Fukushima disaster is a vivid reminder that public opinion is a critical factor in the adoption of energy technologies. Clear-headed analyses of costs, reductions in carbon dioxide emissions and risks are impossible in the shadow of this serious incident; although the communities surrounding the Fukushima plant have been deemed 'safe' by the Japanese government since 2015, only around one in eight of those evacuated have dared to move back to their homes.[34] The popular acceptance and investor confidence needed to build many more reactors will be hard-won. On top of this is the issue of how to treat radioactive waste from nuclear fission power plants, which has a half-life of between tens of years and tens of thousands of years – storing this waste safely is costly and requires almost indefinite maintenance. Despite nuclear fission's potential as a very low-carbon, high-intensity energy source, the International Energy Agency predicts only modest increases in the proportion of power generated by nuclear fission by 2040.[35]

RISING RENEWABLES: SOLAR AND WIND

Leonardo da Vinci might not have been able to implement his plans for capturing and using solar energy, but five centuries later, we certainly can. Today, Italy is one of the world's top producers of solar electricity, both per capita and by total capacity. I was involved in the installation of a particularly spectacular array of solar panels in Palermo in Sicily – with a capacity of 6.8 megawatts, they provide sufficient energy to power at least 7,000 homes.[36] That scheme now looks modest compared to a 2017 installation in Tamil Nadu in India that is around ten times larger.[37] 'I think that solar power has the potential to be the absolute dominant source of energy in the future,' says Richard Friend. On paper, that looks very possible. The total amount of solar energy reaching the Earth each hour matches the amount of energy that the whole world consumes in a year. If we picked a very sunny area, the Sahara Desert for example, and covered a patch about the size of Spain with photovoltaic panels, we could supply every person on the planet with abundant electricity.[38] So far we have barely started to tap this potential, but this is poised to change dramatically. In 1977 photovoltaic electricity generation cost around $80 per watt[39] – at that price, most people dismissed it. Over the following decades, something remarkable happened. 'The cost of solar has followed a trajectory that no one thought it could manage,' says Friend. By 2015, the installed cost of solar had plunged to just $0.30 per watt, around a 250-fold price reduction in less than forty years. Friend says that the key factor forcing this declining cost has been the industrial increase in the fabrication of silicon-based photovoltaics, largely driven by Chinese corporations. Solar is now reaching a tipping point, where the cost of its electricity rivals that produced from coal or gas power plants. While it still only contributes less than 2 per cent to the global electricity supply, that contribution is rising rapidly.[40] Friend thinks the cost of the solar cell is no longer the main limiting factor for the rise of solar energy: 'The silicon in standard cells is now quite a small fraction of the total cost of an installed solar panel,' he says. 'Now the game is all about getting more efficiency out of it.' The best silicon-based solar cells convert only around 25 per cent of sunlight into electricity. If solar panels can become more efficient without getting more expensive, solar power's rise as a major energy source should accelerate dramatically.

Oxford University's Henry Snaith is working on a new approach to photovoltaics that he hopes will soon outperform and undercut silicon cells, and Friend thinks his inventions could be ground-breaking. They demand that we 'put today's inorganic semiconductor textbooks into the dustbin', he tells me. He is working with metal halide perovskite substances, a class of semiconductors that are defined by their distinctive crystalline structure. Like crystalline silicon, these perovskites convert photons of light into electrical power through the photoelectric effect.[41] But, while silicon needs to be extremely pure

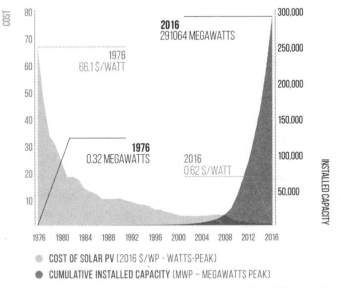

FIGURE 6.10 Watts up, Sun: the cost of solar energy has fallen while its use is increasing.

to work, perovskites can perform well even when shot through with impurities, which makes them significantly easier and cheaper to make. The development of new photovoltaic devices is very different from the complexity, rigidity and expense of the experimental fusion reactor ITER. By experimentation and iterative improvement, the efficiency of perovskite solar panels has been pushed up by a factor of ten in just five years, to 22 per cent.[42] As Friend explains, 'That's very impressive, but there are still a lot of questions that have to be sorted out if they're going to go into massive production, including the toxicity of lead and the durability of materials.' They will have to be better than silicon cells in all respects to overcome the 'power of incumbency', which always slows the penetration of new products into markets dominated by established technologies.[43]

In parallel with the dramatic decrease in the price of solar panels, the cost of harnessing wind energy has also fallen in recent decades. Today's most effective wind turbines have blades the size of the largest commercial airplane's wingspan, and each produces enough electricity to power 7,500 European homes. Wind energy is now being added to electricity grids across the world at an accelerating rate – China installed two wind turbines every hour in 2016.[44] One feature of both wind and solar which hinders their ability to dominate electricity generation is that they are inherently intermittent. The wind blows neither consistently or predictably, and the sun does not shine at night nor penetrate the gloomy weather most common in the industrialised countries of the Northern Hemisphere. As more electricity is obtained from these sources, conventional power plants that use nuclear or fossil fuels are needed to offset the intermittency. These conventional plants are able to increase their output when demand surges. Conversely, when bright sun or strong winds produce electricity that exceeds demand, grids need ways to shed power or reduce generation capacity, in order to avoid dangerous malfunction. Wind and solar farms are often turned off in these situations – nearly a fifth of wind-generating capacity and a third of all solar capacity in China was curtailed in this way in 2016.[45] If wind and solar are to fulfil the potential they have to reduce carbon dioxide emissions, we need to engineer better ways to manage and store the electricity that they produce; this is the principal area of expertise of Nigel Brandon, the Dean of Engineering at Imperial College in London.

THE ENERGY SYSTEM: SOURCES, STORAGE AND COMPLEXITY

'Why do you need storage in an energy system? An energy store is inefficiency … All you do is get less out than you put in,' says Brandon. 'The real value of storage is that it reduces generation and transmission investment. It is an enabler for the system as a whole,' he explains. If storage is available, the infrastructure that is already in place can be used more efficiently. With good engineering, the cost of getting energy in and out of storage is more than offset by the value created by a more efficient and reliable grid.[46] However, the incentives are not yet in place to value storage as a service, so the business case for it is not yet robust; however, that will change in the near future.

Brandon's research is focused on batteries – both improving existing ones and finding new types. The life of the commonly used lithium ion batteries has been dramatically improved, and economies of scale are driving down prices. Brandon, however, points out that while they can be made safe and reliable, he doubts that their cost will ever be sufficiently low to make them the best choice for grid-scale application – at present, he thinks that vanadium flow batteries may be a better option. Unlike lithium ion batteries, vanadium flow batteries store electrical energy in liquid electrolytes.[47] They are non-flammable, do not degrade with repeated charge and discharge cycles, and are relatively cheap to make. Although their energy density is lower than lithium ion batteries, this may not be the most important consideration in a grid-scale storage facility. 'Vanadium flow cells can be engineered with a large volume, and therefore a high capacity. Several countries have recently started to install arrays of them,' says Brandon. Many other new technologies, including liquid metal batteries based on sodium and nickel,[48] metals that are cheaper and more abundant than vanadium, are advancing rapidly.

Batteries are not the only energy storage solution, and nor are they always the best or the cheapest. There are many storage technologies, each of which has properties that are appropriate for different applications. Some store and release power quickly, while others absorb very large amounts of energy and maintain it for long periods of time. The energy grids of the future, especially those with a high proportion of intermittent renewable energy sources, will need a range of these different systems to work efficiently. Brandon believes that surplus electricity could be used to produce hydrogen by electrolysing water.[49] 'You can store a lot of energy if you make it into a chemical bond,' he says. When hydrogen burns, its only by-product is water vapour; replacing natural gas with hydrogen could, therefore, result in large reductions in carbon emissions. It can also be moved around reasonably easily in tankers and by pipelines. Parts of Japan and South Korea already make extensive use of hydrogen for domestic heating.[50] At present, the production of hydrogen is an energy-intensive process, but I am confident that in the coming decades chemical engineers will change that, allowing 'clean' hydrogen to play a role in mid-twenty-first-century energy systems.

Around 85 per cent of all the energy we consumed in 2015 came from coal, oil and natural gas,[51] but there is no doubt that an appropriate response to climate change will change this mix. To provide the energy the world needs at an acceptable cost represents a great economic, engineering and political challenge. With an increasingly large diversity of methods for making, storing and distributing energy, each country and each region needs to find its own distinctive ways to meet both its energy needs and its climate change mitigation targets. They will have to reduce the risk of interruptions to their supply, which are currently caused either by bad engineering or by the disruption to imported supplies that results from geopolitical disturbances. In this complex mix the primary energy sources are nuclear fission, wind, solar, geothermal, hydropower, coal, oil, natural gas and biomass. They need to be stored where they are needed so they are available on demand. Then they need to be converted into

GLOBAL PRIMARY ENERGY CONSUMPTION
(PER CAPITA, PER YEAR)

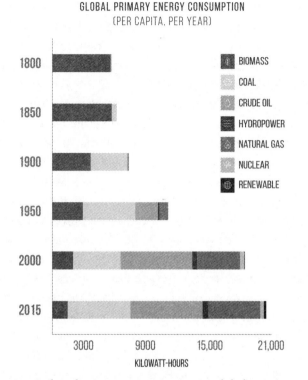

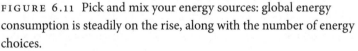

FIGURE 6.11 Pick and mix your energy sources: global energy consumption is steadily on the rise, along with the number of energy choices.

electricity, industrial heat and transportation fuels. And then a mix needs to be created that fulfils the unique needs of individuals, communities or countries. All this amounts to a very large and long-term problem involving cost, security of supply, pollution, climate change and employment. None of this will be effective without the changes in behaviour that result from intelligent regulation and long-lasting incentives. For the past two decades, I have been a consistent advocate of charging for the use of carbon, most probably by imposing a tax; although many have tried, no scheme has yet had a real impact on the behaviour of consumers and investors. Carbon taxes must be designed to force behavioural change, and to do this they need to be at a high enough level to encourage people to change their behaviour – to date, no scheme has been able to achieve this. The implementation of a carbon tax without exemptions, and with measures to penalise countries which do not comply, will require courage and determination to face down vested interests.

As well as the introduction of an effective tax on carbon, the demand for energy can also be managed by the use of technologies that improve efficiency. For example, rather than adjusting the supply of electricity in response to demand, 'smart grids' temporarily reduce the power supply to non-essential devices such as air conditioners, heating systems and refrigerators, until supply and demand is able to rebalance. At times of over-supply, more power is diverted into charging electrical vehicles, storage heaters and other such devices. It is also often forgotten that, in almost every aspect of our lives, we are already achieving more with each unit of energy used – the story of energy is not just one of rising consumption, but also of steadily improving efficiency. Most modern automobiles can travel four or five times further on a gallon of gasoline than the Ford Model T did a century ago. A standard white LED bulb produces light one hundred times more efficiently than a kerosene wick lamp. The first coal-fired power station that Thomas Edison built in New York in 1882 converted less than 2.5 per cent of the chemical energy of coal into electricity; a new combined cycle gas power plant achieves an efficiency of greater than 60 per cent.[52] Furthermore, there is every reason to believe that this rate of improvement in the energy efficiency of our devices, vehicles and systems will continue.

However, as William Stanley Jevons warned over 150 years ago, better efficiency is not an automatic route to reduced consumption. He wrote, 'It is a confusion of ideas to suppose that the economical use of fuel is equivalent to diminished consumption. The very contrary is the truth.' As he predicted, efficient steam engines led to the widespread use of steam power and a growing demand for coal; total energy consumption grew in step with economic growth, with carbon dioxide emissions rising in parallel. Recently, however, there have been encouraging signs that the pattern may have shifted.

Between 2014 and 2017, global carbon dioxide emissions plateaued, while overall economic activity continued to rise[53] – perhaps this is confirmation that the tight link between energy consumption, carbon dioxide emissions and economic growth is breaking. Part of this success is surely attributable to the considerable improvements in energy efficiency that are being implemented in many economies and devices around the world. Now carbon dioxide emissions need to fall.

Most climate scientists agree that, if we are to avoid the worst effects of climate change, extensive de-carbonisation of our energy systems needs to occur by the middle of this century. As the cost of renewables plummets, new storage technologies become available and ways to regulate demand and efficiency improve, it is tempting to believe that this ambitious target is within reach. 'Deep de-carbonisation of the electricity sector; I can see how to do that,' says Moniz. However, electricity will not power everything and, in his view, 'the idea of fully electrifying the transportation industry is out of the question'. Large commercial trucks, cargo ships and aircraft that could be powered by electricity have yet to be engineered. Furthermore, many industrial processes require the heat that is directly generated from fossil fuels. 'It doesn't violate the laws of physics to generate that heat with electricity, but that isn't the way it's going to work,' Moniz concludes. Even if it became possible to electrify all transport and all heat sources, national electricity grids would have to be entirely re-engineered to serve the massive increase in demand. While carbon dioxide emissions must be greatly reduced, we should not set ourselves the target of eliminating fossil fuel use entirely. At the moment, such a target defies credibility and will be irresponsible if it serves to slow down economic development. Fossil fuels are too powerful, too convenient and too abundant to abandon entirely. If we can engineer cost-effective ways to manage, store or use the carbon dioxide that burning these fuels produces, Moniz suggests 'you can give fossil fuels a break'. Some may see this as a delaying tactic of an incumbent hydrocarbon lobby, but I see it as an urgent priority.

Plants and algae have been converting carbon dioxide and water into fuel for nearly 3 billion years, but most plants convert only 3 to 6 per cent of the solar energy that they absorb into biomass. For plants to absorb significant amounts of carbon dioxide from the atmosphere, impractically large amounts of land would need to be planted and kept alive in perpetuity.[54] Chemists, biologists and engineers are scrutinising the molecular basis of natural photosynthesis, hoping to improve its efficiency. Others want to emulate or improve what plants do, with the use of chemical catalysts. In fact, catalysts that destroy carbon dioxide by splitting carbon from oxygen already exist,[55] but they are expensive and demand a prohibitively high input of energy to work. Ideal solutions would not just destroy carbon dioxide, or capture it and bury it, but would instead turn it into a valuable commodity. One promising scheme would harvest carbon dioxide from industrial sources and combine it with hydrogen, creating synthetic methane, methanol or ethanol. These are useful fuels in their own right, but also important feedstocks that can be used to produce denser fuels and petrochemical-derived products such as plastics. In ideal circumstances, the energy needed to create these fuels would come from sunlight, which would generate electricity to split water into hydrogen and oxygen, and heat to react hydrogen with carbon dioxide. This has been demonstrated in several laboratories around the world, but a great deal of engineering will be needed before these processes will work economically.[56] There are so many promising schemes in development that I have no doubt that carbon capture and storage or use will play a significant role, but regulatory acceptance and incentives are required to make it an attractive economic, as well as ecological, proposition.

A TIME FOR ACTION
Stood at the lectern at Stanford University back in 1997, I decided to meet the challenge of climate change head-on. I acknowledged that the fossil fuel industry was at the centre of the problem, but also described how it could be part of the solution; the head of the

American Petroleum Institute said afterwards that my speech meant I had 'left the church'.

In June 2018, at a unique summit hosted by Pope Francis, I find myself back in the church. Assembled beside me are the leaders of many of the world's biggest oil and gas companies, investors who oversee trillions of dollars' worth of capital and many of the energy sector's leading thinkers and policymakers, along with various priests and bishops of the Catholic Church. We are here to talk about climate change and how we can tackle it without curtailing the economic growth of the world.

The serene sixteenth-century chapels and palaces of the Vatican lend the appropriate gravitas to proceedings. The tone is one of humility, and the discussion is marked by a pragmatic optimism. Finally, more than two decades after I first urged the hydrocarbon sector to take heed of the grave reality of climate change, the terms of the debate have shifted – what I hear at the Vatican reaffirms my conviction that we now have the pieces we need to solve this most pressing and complex of global puzzles.

We now have at our disposal the means to provide more energy to more people, while simultaneously extracting a lower economic, humanitarian and environmental cost. However, putting the right technology in place quickly will be expensive and will require that we have the courage to confront vested interests and design effective incentives, in order to create a market pull in the right direction. History shows us that energy transitions, from the use of only our muscles to wood to fossil fuels and beyond, are generally slow and incremental. The current transition is unlikely to be an exception, but this time we may not have the luxury of waiting to see what unfolds. We need to keep inventing better and more cost-effective energy sources, more efficient machines and more intelligent ways to link our infrastructure together. As Ernie Moniz makes clear, 'we need to innovate vigorously in all of these areas'. In the end, innovation gives us options – the challenge is to combine them in the right way, with the right incentives, for the benefit of humankind. As Pope Francis summarised at the Vatican summit, 'Civilisation requires energy, but energy use must not destroy civilisation.'

As the world grows, we will need increasing amounts of energy. Even if the US and Europe become much more energy efficient, the rest of the world will consume more, as they strive to match our standard of living. This will only be possible with the use of fossil fuels, with efficient ways to capture and store or use carbon dioxide, in combination with the greater application of zero carbon sources such as renewables, as well as improved versions of nuclear fission. As things stand, nuclear fusion remains a distant prospect and is not the magical solution to all our energy needs. There are many forecasts regarding the future mix of energy sources, the cost of implementing the mix and the concentration of carbon dioxide in the atmosphere that will result. There remains a great deal of uncertainty, but almost all these forecasts indicate that the world must immediately increase its investment in energy by at least 60 per cent above the level seen in 2015. The amount of hydrocarbons in the mix will depend on effective

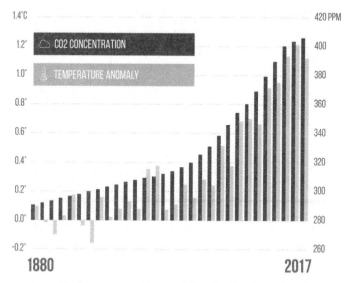

FIGURE 6.12 No heart-warming trend: levels of carbon dioxide rising in tandem with global temperatures.

incentives to reduce the use of carbon, and continued engineering and commercial breakthroughs in the production of all other forms of energy.[57] How we set in motion these actions to provide all of human-kind with the energy we need to progress, while also sustaining the biosphere's capability to nurture human life, is among the most profoundly consequential decisions that we will have to make.

7

Move

I was fortunate enough to travel on Concorde many times, and each journey was a thrill. I remember a rare vacation trip to Barbados when the plane travelled at twice the speed of sound for almost the entire journey, because it was passing over water and its fearsomely noisy jet engines would not affect anyone on the ground. Then there were many business trips from London to New York when, because it flew faster than the Earth turned on its axis, you would land an hour before you had left. You could do a full day's business in New York and come home on the overnight flight – something I did regularly for almost twenty years. In the late 1990s, it made possible the time-sensitive negotiations to create what was then the world's largest merger of two oil companies, BP and Amoco. I often met my American counterparts at nondescript hotels close to New York's JFK Airport, having arrived in a way that was not only time-efficient, but also retained the glamour of the air travel of the past.

Concorde was the world's only technically successful supersonic passenger aeroplane. It provided a glimpse of the future, and yet it was a technological marvel built from 1950s parts. Today's planes can take off, land and navigate with minimal human input, but Concorde was a very different kind of aircraft. I remember visiting the flight deck and feeling like I was stepping back into the past. Analogue dials, switches and Bakelite knobs covered every available surface.

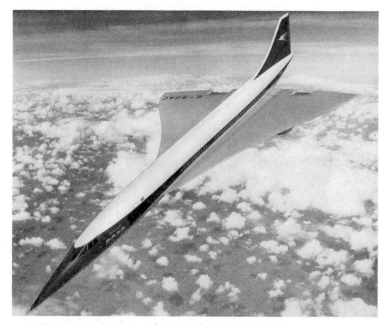

FIGURE 7.1 Supersonic travel took off with Concorde in 1976, but was silenced in 2003 following a fatal crash.

For its time, Concorde's fly-by-wire technology, the precursor to today's autopilot systems, was advanced, but it always required a vigilant flight crew of three; sitting behind the pilot and co-pilot was the flight engineer, who had the particularly critical role, now entirely computerised, of maintaining the plane's centre of gravity, by constantly pumping fuel between thirteen separate tanks. He also had to pay close attention to the air conditioning systems, because flying at twice the speed of sound caused such great friction and compression of air that the aircraft would heat up dramatically. I can still remember the cabin getting noticeably warmer during a flight.[1]

For two and a half decades, Concorde challenged the world's preconceptions about the possibilities of travel. Unfortunately her distinctive take-off no longer rattles windows beneath her[2] – the last Concorde flight was in November 2003. Opinion is still divided as to whether this aircraft was decades ahead of its time, or a fuel-hungry relic from a time when oil was cheap and state sponsorship allowed engineering projects that were economically unviable. No commercial airliner has breached the sound barrier since. In fact, air travel is now almost always slower than it was when Concorde first flew; then, for example, a typical flight from New York City to Houston, Texas, would have taken less than three hours, while it now takes at least three and a half. This is because airports and air lanes are more congested, and modern turbofan jet engines work most efficiently below their potential maximum speed.

Although air travel has stopped getting faster, progress in the aviation industry has not slowed – rather, engineering has opened the opportunity of air travel to a huge sector of the global population. One hundred and twenty years after the eminent UK physicist and engineer Lord Kelvin stated that 'no balloon and no airplane will ever be practically successful',[3] flying is a routine and safe feature of everyday life. But, rather than becoming steadily bigger and faster, the trend is now towards lighter, quieter and more fuel-efficient commercial aircraft. Flying may no longer have the mystique or the speed of Concorde, but the advantages of moving people and goods rapidly across the globe are no longer the preserve of the elite.

At any moment, more than a million people are travelling on an airplane. Three-and-a-half-billion passengers, the equivalent of half the world's population, board a flight each year, and our crowded airport terminals reveal our hunger for air travel. Furthermore, just as our airways are becoming congested with flights, the streets of our cities, from Mumbai to Mogadishu to Manhattan, are increasingly busy with cars – in the US, there are now more cars than registered drivers.[4] Cruise ships also attract an increasing number of customers every year and, from high-speed intercity links to commuter trains, railways rarely seem able to satisfy demand.

Why is our appetite for travel so persistent? Partly it is a necessity – as a result of improved transport, trade between nations has grown steadily and now accounts for nearly two thirds of global GDP.[5] Countries rely on each other and, in business, eye contact and the old-fashioned handshake still have a resonance that even

the most cutting-edge virtual reality environments cannot replicate. People still need to travel to do their jobs, even with modern
computers and communications networks, which actually appear to
increase rather than reduce the demand for face-to-face meetings.
We are also travelling more because we can afford to do so; with more
efficient vehicles and infrastructure, both short- and long-distance
travel has become dramatically cheaper. However, the persistence of
physical travel goes far beyond practical need and affordability – we
have a deep-seated and restless curiosity about the world. No matter
how much we read about other countries, watch them on our screens
or consume their products, we still yearn for the chance to taste, hear,
smell and feel the world for ourselves. To do this, we willingly put up
with the modern-day discomforts of airport security, jet lag and lost
baggage. It is this deep urge to keep moving and to constantly explore
that has driven progress towards the increasingly prosperous and free
world that we inhabit today. We must have possessed our appetite
for exploration from our beginning – what else could have drawn
our distant ancestors to a voyage across the Timor Sea to Australia
65,000 years ago,[6] when they first left the Afro-Eurasian landmass?
Their curiosity carried them to new lands, which they would eventually call home.

TRADE, DISEASE AND POWER
Over millennia, trade gradually spread between Europe, the Middle
East and North Africa, with raw materials and goods being transported
by sea and land. The Romans later built a network of roads through
the whole of their empire, and later still the Silk Route reached all the
way to China in the east and to the Indian subcontinent in the south-
east. These links made it possible not only to trade valuable items but
also to exchange stories and ideas, which in turn encouraged cultural
development and helped to erode people's natural fear of foreigners.
Other, less welcome, things were also exchanged. The bacteria that
caused the Black Death, which killed half of all Asians and Europeans
during the fourteenth century, started in China and spread along
trade routes.[7] Christopher Columbus took smallpox, measles and

FIGURE 7.2 A print with a view: shipbuilding in Venice, from
Breydenbach's *Peregrinatio in Terram Sanctam* (1490), the first illustrated
travel guide.

influenza to the Americas, and brought syphilis and tuberculosis back
with him. And today, our transport systems can spread disease more
easily than ever before. In 2003, for example, SARS first appeared in
mainland China and then spread to thirty-seven different countries
within weeks.

Just as in the time of Alexander the Great and the Silk Routes, our
transport systems are still used to spread war and terror. The arteries
that connect nations are critical for trade and exchange, but they are
also powerful ways to gain influence and power. The search for political as well as commercial influence is behind China's 'Belt and Road'
initiative that many call 'the modern Silk Road', a hugely ambitious
plan to lay down roads, railways, pipelines and new ports throughout
much of Eurasia and Africa. Hundreds of billions of US dollars' worth
of projects are planned or underway, aimed at accelerating trade with
nations along its route and also those nations' development.

I own a copy of an intriguing fifteenth-century book by Bernhard von Breydenbach, a canon of Mainz Cathedral in Germany, who, in 1483, set off on a pilgrimage to the Holy Land. Unusually for the time, Breydenbach took an artist, Erhard Reuwich, with him to record the towns and people they encountered on their travels. The book that they put together on their return to Germany is the first illustrated travel book.

Reuwich's exquisite wood engravings are impressively realistic and full of life. A five-foot-wide panoramic view of Venice, which includes accurate renderings of many of the city's public buildings, and evidence of shipbuilding in the Arsenale (which is discussed in Chapter 8), folds out from the book. The text is a strange blend of styles; in parts, Breydenbach voices his disapproval about the habits of the 'infidel' Muslims, Jews, Arabs and Turks he meets, which is what his devout Christian sponsors back home would have expected of him. Many of his criticisms, however, are offset by his obvious fascination not only with the differences between these groups of people but also with his appreciation of their shared humanity. For example, he describes how, while crossing the desert, the people they met went to 'great trouble' to draw water from the ground for the thirsty travellers, 'for God's reward'. And there is no attempt in Reuwich's engravings to belittle the Saracens, Mamelukes and other people they encountered. When I first read the book, I felt I was seeing the birth of a more inclusive and tolerant world view, after the atrocities committed during previous centuries by both the Christian crusaders and their enemies. This exposure to and acceptance of difference made possible the progress towards the more cosmopolitan and open-minded future that we live in today.

FROM HORSE POWER TO STEAM POWER

The modes of transport available through most of our history were slow by today's standards, but they still enabled us to shape the world. The domestication of the horse was the first advance that accelerated people's ability to travel over land. Humans had long hunted horses for their meat, but the first convincing evidence of their domestication dates from the middle of the fourth

FIGURE 7.3 A disruptive invention that stirred up ways of travel, warfare and dissemination of information: the stirrup (eleventh century).

millennium BC, and is associated with the Botai culture in northern Kazakhstan. Analysis of some of the 300,000 horse bones found at one of the key archaeological sites there showed distinctive patterns of wear on the horses' teeth and lower jaws, associated with the use of a bit and harness. Fragments of pottery contain traces of fats from horse milk, which is evidence that at least some of these tamed horses were being ridden and milked more than 5,000 years ago – as archaeologist Robin Bendrey notes, 'You can't milk a wild horse.'

At the British Museum in London, a curator shows me an example of an invention that entirely changed the experience of riding on horseback. From its origins in the Jin dynasty of China (AD 265–420),[8] the stirrup spread throughout the world. At first it seems so simple and intuitive as to scarcely qualify as an invention: a simple hoop of iron with a flat base to hold the sole of the foot, and a fastening at the top to attach to a saddle. The stirrup in front of me at the British Museum dates to the tenth or eleventh century and was found, a thousand years later, in the River Witham in the east of England.

The stirrup transformed the rider's relationship with the horse. Firstly, it made it possible to endure much longer journeys in the saddle, and secondly, it gave the rider more stability and control. The Bayeux Tapestry depicts William the Conqueror's Norman cavalrymen using stirrups during their invasion of Britain in 1066. With their feet stabilised, their arms were free to hold long shields with which they could defend themselves and attack their opponents with more force and accuracy. Such dexterity eventually enabled them to break the defenders' infantry line and rout the Anglo-Saxon army, killing King Harold in the process.

So England came under Norman rule thanks, at least in part, to the stirrup. By enabling the creation of military forces which were composed of heavily armed knights and horses, it was also a factor that enabled the rise of the feudal system of governance, which dominated most of Europe during the medieval period.[9] Further east, it also helped the mounted hordes of Genghis Khan's army to establish the largest contiguous empire that the world has ever seen. The impact of the stirrup was as important as the introduction of the wheel; efficient travel on horseback did much more than simply enable people to fight more efficiently – it also connected them, allowing them to cover longer distances over shorter time spans. And it was not just people and wares, but also information that could move with greater ease. During the fifth century BC, mounted couriers could carry messages along a 2,700-kilometre section of the Persian Empire's Royal Road in just seven days, eighty-three fewer than the same journey would take on foot. 'There is nothing in the world that travels faster than these Persian couriers,' the Greek historian Herodotus wrote. Such rapid dispersal of ideas was a prerequisite for innovation and the expansion of civilisation.

Even in the rapidly industrialising West, horses remained a major source of transport until well into the twentieth century, long after the invention and refinement of the steam engine and the internal combustion engine. The number of horses in the US, for example, peaked as late as 1915, when there were more than 26 million of them, or about one for every four people.[10] The gradual and uneven transition away from animals and towards mechanical power is a clear illustration that technological development is rarely orderly or predictable, because it is subject to many social, political and economic forces. As David Rooney, formerly Keeper of Technologies and Engineering at London's Science Museum, explains, 'No technology is inevitable … history shows you get winners and losers for all sorts of reasons.' It is also clear that new technologies almost never replace old ones completely, but rather are added to them and continue to coexist, often for decades. 'That's not because people are conservative, lazy or doggedly sticking with the past. It's because those [existing] technologies work, they are appropriate and people have invested in the networks that maintain them,' Rooney continues. Thus, while horse-drawn vehicles for passenger transport were replaced early in industrialised countries, horse-drawn freight persisted much longer. 'If you had a fleet of twenty or thirty horse-drawn freight wagons in London, it was a massive investment to replace them with diesel trucks,' says Rooney. In the history of transportation, similar dynamics play out time and again – this, I believe, is an important point to remember as we listen to today's excitable forecasts of fast-approaching transport revolutions.

The turn of the nineteenth century was a period of rapid innovation – the steam engine, the power loom, the bicycle, gaslight and the hot air balloon all arrived during this period. The phrase 'the March of Intellect' came to stand for the growing confidence among scientists and engineers that their work could drive progress towards a better world for all. A wonderful 1828 engraving by William Heath captures both the spirit of the age and the newfound willingness to imagine a technologically enabled future. In it, Heath pictures an array of extraordinary vehicles: a steam-powered horse charges

FIGURE 7.4 Transport Uber Everything: William Heath's satirical vision of future transportation (1829).

through the foreground; cannons are used for 'quick conveyance for the Irish emigrants'; and, in the sky, a mechanical bat takes convicts to New South Wales in Australia. Heath even presages today's speculative 'Hyperloop'[11] technology, a proposed transport system whereby a train or separate 'pods' would travel at high speeds through partially evacuated tubes or tunnels. In Heath's vision, a cast-iron cylinder of the 'Grand Vacuum Tube Company' takes passengers directly to Bengal, through depressurised tubes at great speed.

The most visible sign of early nineteenth-century technological progress was the rapid growth of railways in the UK. After the first line between Stockton and Darlington had been opened in 1825, dozens of railroad companies were formed, aggressively promising a virtually risk-free investment. Money poured in and tracks were laid at an accelerating rate. Andrew Adonis, the former UK Secretary of State for Transport, describes the phenomenal speed at which railways grew: 'If you look at a map of Britain's railway network in

1850, it's not much different from now. Almost the entirety of the railway system we now have was planned and built in twenty years.'

At first, the appearance of these 'iron roads' and the noisy, smoke-belching steam engines that travelled on them created a backlash. Unproven locomotive technology and tracks that had been built in haste meant that accidents were common. Farmers worried that cows in nearby fields would miscarry or stop producing milk, and that sheep's fleeces would be blackened by the smoke. Even eminent scientists suggested that high-speed travel on a jolting train could cause problems with vision, breathing and mental health. But, as Adonis explains, these fears were short-lived: 'Looking at the history of it, the thing that strikes me most clearly is how quickly things moved on from this being a dangerous and outlandish technology to being generally accepted.'

However, despite improving safety and the clear advantages of train travel, the 'Railway Mania' of the 1840s was unsustainable. Over-zealous speculation had created an investment bubble, equivalent to the 'dotcom' bubble of investment in the World Wide Web in the late 1990s. A crash was inevitable and when it came, many private investors were ruined, several railway companies collapsed and planned routes were never built. It did, however, leave the country with the most advanced and extensive railway network in the world, a system that would soon change almost every aspect of contemporary life. Factory output increased dramatically. Working families could visit the seaside. Fresh milk could reliably reach London before it soured. But, of course, not everyone agreed that rail travel represented beneficial progress – John Ruskin, for example, declared railways 'the loathsomest form of devilry now extant ... destructive of all wise social habit and natural beauty'. However, there can be little doubt that, by the end of the nineteenth century, railways were helping to create a more connected and prosperous world.

Railways are still a crucial part of transport infrastructure in most parts of the world, but the ways in which different countries use and invest in them vary enormously. Adonis explains how Japan constructed the first ever high-speed rail link, the Shinkansen 'bullet train' between Tokyo and Osaka; soon after the line opened, in 1965, trains were travelling at over 200 kilometres per hour, which halved the journey time between the cities. Adonis points out that the UK is only now beginning construction of a high-speed rail link between its two largest urban centres, London and Birmingham: 'That's two whole generations,' he exclaims incredulously. In the US there is still only one stretch of track, between Washington DC and Boston, on which trains can reach the same speed that the first Shinkansen managed six decades ago.

In Japan, meanwhile, trains routinely travel at 320 kilometres per hour, while a magnetic levitation train there recently hit 600 kilometres per hour on a test track. 'As a historian, I think this idea that everything is speeded up and faster in the modern age ... and that state-of-the-art technology inevitably sweeps across the world is complete and utter balderdash,' says Adonis. He goes on to describe how the early decision to invest in high-speed rail involved a good measure of happenstance. The dominant thinking in Japan in the 1950s was that railways were an outdated technology that was unable to compete with the flexibility of automobiles. The chief engineer at the Central Japan Railway repeatedly made the logical argument that, as Japan's cities continued to grow, so would the challenge of providing rapid mass transit. Adonis summarises the engineer's position: 'As soon as automobiles became ubiquitous, they would cease to be rapid, because it wasn't going to be possible to accommodate them [on the road system]. Whereas railways, if you went to the next level of engineering, could be both more ubiquitous and more rapid.' However, this sound reasoning was not the factor that won the argument – what swayed the politicians was the upcoming 1964 Olympic Games in Tokyo. Japan's rapid new railway was seen as a powerful symbol of the country's post-war emergence as a major industrial and technological power.[12]

As the steam engine was transforming travel on dry land, it also completely and irreversibly revolutionised maritime travel. For the previous three centuries, the age of sail had steadily solved the enduring mysteries about what might lie over the horizon – in

Chapter 8 I will return to some of the extraordinary shipbuilding and navigational innovations that opened the world in this way. But it was only with the widespread adoption of the steamship during the second half of the nineteenth century that the world's first era of globalisation began. Before steam, trade routes and shipping times were dependent upon currents and the prevailing winds. That changed abruptly, as new routes opened and freight costs began to fall dramatically.[13] Emigration also became a more viable option for many; as a result, the world entered a new age of rising integration and interdependence that would only be derailed by two world wars and the Great Depression.

Many steamships remained in service until the 1950s, though by then all new ships were propelled by much more powerful and efficient diesel engines. Today, vast container ships that are up to 400 metres long, with the ability to carry hundreds of thousands of tons of cargo, ply the world's shipping lanes. In major ports, highly automated systems choreograph the loading and unloading of tens of thousands of six-metre-long shipping containers from each of these mighty ships. It is these humble, standardised steel boxes that move the vast bulk of traded goods, from perishable foodstuffs to tablet computers and almost everything in between, around the globe, dependably and affordably. Thanks to this convergence of innovations, the world's economy took another great leap forward.

TAKING FLIGHT

Our vision of the future is usually shaped by the things we see around us. When it came to conceiving machines that could fly, Leonardo da Vinci based his fifteenth-century plans for a 'helicopter' on the spiralling motion of a seed falling from a maple tree. In 1893, the German aviation enthusiast Otto Lilienthal built a glider that could keep him airborne for 250 metres, as he flew from the top of a small hill. He was convinced that a more successful flight would require a machine with wings that would flap like those of a bird. Lilienthal's quest for birdlike flight eventually killed him, when he crashed and broke his neck after one of his experimental flying machines lost lift.

Engineers often make most progress when they forget their preconceptions about the way the world works and start again from first principles. The Wright brothers won the race to build the first powered heavier-than-air aircraft because they realised that Bernoulli's principle[14] worked, which meant they did not need to replicate the flying motion of a bird. While their contemporaries were obsessed with building devices that were as stable as possible, the Wright brothers realised that the critical issue was control; if an airplane could be steered effectively and responsively in all three dimensions, the pilot could accommodate buffeting by turbulence and keep the machine in the air.[15] The Wrights' first flight, on 17 December 1903, covered just thirty-seven metres, but it represented a significant step for humankind, since it broke an important barrier and unleashed rapid progress. The first commercial passenger flight, covering the thirty-five-kilometre gap across the bay that separates St Petersburg and Tampa in Florida, took place within eleven years. The two world wars that followed ensured that planes matured into a more robust and efficient form of transport.

In 1957, when I was nine years old, my mother and I moved to Iran to join my father, who was working in the oilfields there. Two years later, my parents decided to send me to boarding school in Ely in the UK. Thoroughly enjoying the colour, heat and vibrancy of Iran, I was in no hurry to return to cold, grey reality at the end of each holiday, but one consolation was always the exciting prospect of the flight back to Europe. At first, the propeller-driven planes had to make several stops on the way, hopping from Abadan, to Beirut, to Rome, and then to London. But during this time, jet travel came of age.

The gas turbine, the innovation that made jet-propelled aircraft possible, is the only type of engine in use today that was invented during the twentieth century. In 1928, the Royal Air Force pilot and engineer Frank Whittle drew up the first viable designs for a jet engine. I met Ralph Robins, one of the pioneers of jet engine engineering, to discuss the impact it has had on our world; he began his career as a graduate apprentice at Rolls-Royce in 1955, before rising steadily through the company, eventually becoming chief executive and then chairman. Although he has been retired for over a decade, Robins is as energetic as ever and well-versed in today's latest developments. He does, however, admit to deciding this year to consign his 1938 Delahaye racing car to the garage and retire from competitive motor sport: 'I was, I think, Europe's oldest racing driver,' he says with a modest smile. Over the course of Robins' career, he has helped to transform air travel from an expensive and unreliable option for the few, into a ubiquitous and affordable mover of people and freight for the many. 'It's all been due to the gas turbine,' he declares. 'Whittle's work has affected more people than any other engineering invention in the world.'

Despite conceiving the idea and building the first working engine, Whittle was not the first person to build an airplane that flew solely under jet power; that was achieved in 1939 by engineers in Germany. Robins explains that Whittle's idea was 'pooh-poohed by the Air Ministry', which dramatically slowed down his work.[16] While he

FIGURE 7.5 Comet with wings: the first commercial jet liner, the de Havilland Comet (1952).

had to scrabble around to get private funding, German turbine engineers, led by Hans von Ohain, moved steadily forward, at their government's expense. At Whittle's memorial service, Robins met von Ohain, who 'told me something that really worried me. He said, "If Whittle had had the British government's support that I had from the German government, the UK would have had a jet fighter by 1939, and the whole history of the Second World War would have been different."' In any event, by the end of the war, Germany, Britain, the US and Japan had each developed capable jet fighter aircraft.

The first commercial passenger jet followed in 1952 – the de Havilland Comet was an elegant airplane with four powerful jet engines buried within its swept-back wings. The promise of this new form of transport was obvious: it flew faster and higher than any earlier commercial airplane. Today's airliners are scarcely faster than the original Comet, but their engines are dramatically more fuel-efficient; that has been a key factor in driving down the cost of air travel. Robins played an important part in this revolution. His first assignment at Rolls-Royce was to work on a new type of jet engine, called a 'bypass engine', in which a proportion of the air entering the jet is passed around the engine's main core to improve its efficiency. Since then, the diameter of jet engines has grown steadily wider as their 'bypass ratio' has increased and, as a result, their efficiency has improved.[17]

As well as now using much less fuel per kilometre travelled, jet travel has also become very much safer. The first Comet was a work in progress. During 1953 and 1954, three of these aircraft broke up soon after take-off, killing everyone on board. The Comet's engineers had not appreciated the immense strain placed on the fuselage by repeated pressurisation and depressurisation and, as a result, small cracks had appeared throughout the metal airframes, especially around the doors and windows, allowing the aircraft to be torn apart at the seams. These disastrous events led to construction techniques improving and safety being tightened up, and also gave birth to the

FATAL AVIATION ACCIDENTS PER MILLION COMMERCIAL FLIGHTS

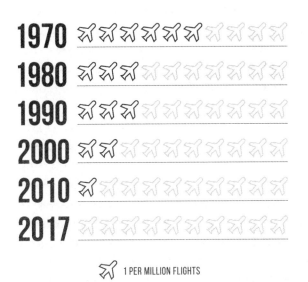

1 PER MILLION FLIGHTS

FIGURE 7.6 A safety briefing for your attention: more flights, fewer fatalities.

fourth iteration of the Comet. This safe airplane was the one that most often carried me to and from Iran during the 1960s.

Safety continued to improve with the development of the 'fly-by-wire' system, the precursor of the autopilot, coupled with better air traffic control procedures. 'I'm sure you're aware that most aircraft fly themselves these days,' says automation expert Hugh Durrant-Whyte, 'because pilots aren't good enough.' He is joking, but there is an element of truth in his remark. By most statistical measures, flying is now the safest way to travel – between 2000 and 2009, there was an average of just one death for 14 billion passenger miles flown in the US.[18] In fact, if you took an 800-kilometre flight every day for a year, your risk of death would still only be 1 in 85,000. Given the complexity of modern aircraft, the extreme physical stress of flight and the congestion of many of today's air lanes, these safety figures are a triumph of engineering.

By contrast, though it may not feel that way, travelling by road is much more dangerous than flying. Road accidents claimed 1.3 million lives in 2015, making it the only thing in the World Health Organization's list of the ten most common reasons for death that is not directly related to disease.[19] Statistically speaking, automobiles are the most dangerous machines that we interact with daily.

DRIVING: A SAFER AND CLEANER FUTURE

Automobiles are not inherently unsafe. Today's vehicles are equipped with an array of effective safety features. From the simple seatbelt to sophisticated electronic stability control, adaptive cruise control and crumple zones, automobiles are better than ever at avoiding crashes and preventing passenger injuries. The problem, however, is the driver; detailed analysis of thousands of road accidents in the US pinpoints driver error as the cause of 94 per cent of all incidents.[20] The best way to improve road safety, therefore, would be to do away with the driver entirely; robotic automobiles, which perceive the world through arrays of electronic sensors, learn from one another and use artificial intelligence algorithms to make better-than-human judgements will, at some stage, be able to drive more safely. However, we are not there yet.

Safety concerns are not the only reason why a reappraisal of the automobile is worthwhile. If life in the nineteenth century was turned on its head by the arrival of rail travel, the automobile had an equally significant impact in the twentieth century. It has shaped the layout of cities all over the world, expanded suburbs, claimed space for new highways, motorways and parking lots, and created the out-of-town shopping centre. Nowhere is this more obvious than in the US, where many of the major cities grew in parallel with the extraordinary boom in the country's automotive industry. The automobile gave its owner freedom and status. Demand rose rapidly throughout the twentieth century, which stimulated the expansion of the oil industry. In addition, producers of steel, aluminium, electronics, rubber and plate glass all expanded their output to serve the automotive industry. More and more people want automobiles;

global production has increased by around 2 million units every year during the last decade.[21] Automobiles and other light vehicles[22] presently consume about 30 per cent of the crude oil produced in the world,[23] but that will decline as they become more efficient and many become powered by electric motors.[24] Policymakers are taking aim at the pollution that automobiles emit, the space they occupy and the congestion they create. A recent analysis of road congestion in France, Germany, the US and the UK estimated that productive time wasted, excess fuel burned and rising delivery charges costs more than $200 billion each year.[25]

At the beginning of the twentieth century, the issue of how best to power an automobile had still not been resolved, and the same question is again being debated. It was not at all clear at the dawn of the automobile age that gasoline would come to dominate; in 1901 some 60 per cent of automobiles in America were electric, while the rest ran on gasoline or steam.[26] Electricity represented the future, and at first it looked like the most promising solution. The first car to

FIGURE 7.7 Pure electric: 60mph on battery power, *La Jamais Contente* (1899).

drive faster than sixty miles per hour was electric,[27] and in 1911 the *New York Times* declared, somewhat prophetically, that 'electric cars are ideal, for they are cleaner, quieter, and much more economical than gasoline-powered cars'.[28] However, gasoline won out for a variety of technological and sociological reasons. The first electric cars, although suited to well-paved city streets, struggled to master hilly or rough terrain, and their batteries limited their geographical range. The ways that different engines were marketed also had a big influence. Gasoline cars were associated with heroic masculinity – they were difficult to drive, since the driver had to conquer the gearbox, and also because they appeared more powerful. As a result, many men refused to purchase electric vehicles, which became associated with weakness. We learn from this, once again, that pure rationality is never the sole explanation for technological transitions.

The current domination of the market by gasoline and diesel-powered automobiles means that road vehicles are a major source of carbon dioxide emissions. Nearly one quarter of carbon dioxide emissions from burning fuel come from transport, and emissions from road vehicles are three times greater than emissions from ships, aeroplanes and trains combined.[29] The efficiency of new automobile engines in the US has doubled since 1975, and continues to rise,[30] but this has still not been enough to curb carbon dioxide emissions – such gains help only if people drive less, but instead they are driving more.[31] To make a substantial dent in greenhouse gas emissions, we need to make big changes to the way we fuel and power our road vehicles.

Carbon dioxide is not the only by-product of the internal combustion engine that causes damage; these engines also emit complex mixtures of toxic fumes, including unburned hydrocarbons, carbon monoxide, oxides of sulphur and nitrogen and microscopic particulates, which are known to cause cancer and heart and breathing problems. In Western Europe, air pollution from road transport is responsible for nearly as many deaths as road accidents[32] – making engines cleaner is thus clearly an important and achievable way to improve quality of life. As Andrew Adonis says, 'Now that it's clear that diesel is a massive health hazard, the push to electrify road transport is going to accelerate. I think that will happen very swiftly.' Electric motors do not produce any polluted air, but they currently power just 0.2 per cent of the cars on the road today.[33] As Nigel Brandon, the Dean of Engineering at Imperial College in London explains, 'the electric motor drivetrain[34] is three to four times more efficient than an internal combustion engine'. Up to 80 per cent of the energy stored in a battery can be converted into useful kinetic energy, whereas only between a fifth and a third of the energy stored in gasoline or diesel reaches the wheels.[35] Gasoline, however, has a massive advantage in energy density terms; the amount of energy in just a cup of it far exceeds that in a very large battery.

Brandon sees the electrification of light vehicle transport as inevitable. Dieter Zetsche, Chairman of Daimler AG and Head of Mercedes-Benz Cars, fully agrees 'that ultimately we need zero carbon', though adds that, 'whilst an all-electric future has some likelihood, it's too early to call that a fact'. Although the range of battery electric automobiles is increasing, the fact that the capacity of batteries limits the distance that can be driven between charges remains a key concern.[36] Zetsche also points to the physical challenge of carrying around a battery that can weigh up to 600 kilograms, and foresees great challenges associated with sourcing the raw materials required for a massive increase in lithium ion battery manufacture. Beyond the automobile, the prospect of creating batteries suitable for heavy-duty transport, including large goods trucks, cargo ships and airplanes, presents engineers with even more significant challenges. 'It's a power-to-weight ratio problem,' explains Ann Dowling, President of the Royal Academy of Engineering. The energy density of batteries is still far lower than that of liquid fossil fuels, so driving large vehicles over long distances requires prohibitively heavy batteries. 'Airplanes will not be electric,' says Brandon. 'It's pointless. You'd end up just carrying your own batteries and nothing else.' Ralph Robins agrees, pointing out that 'The gas turbine is pretty difficult to beat for aviation purposes.' However, Dowling is not so sure and thinks that hybrid planes may soon be a reality – these would use conventional

jet engines for take-off, before switching to arrays of smaller, electrically driven propellers during high altitude cruising.

Batteries are not the only option for powering electric motors; fuel
cells efficiently convert the chemical energy stored in a fuel – most
commonly compressed hydrogen, natural gas or ethanol – into electricity. Unlike batteries, which take time to charge, fuel cell-powered
vehicles can be refuelled rapidly. They could be a good solution for
powering heavy trucks, and some Japanese manufacturers have
already released hydrogen fuel cell automobiles onto the market.
Their range beats that of the best battery-powered electric vehicles,
but, unless you live in Japan, where a hydrogen infrastructure is being
put in place, the utility of such vehicles will be very limited. At the
time of writing, the UK has only one hydrogen pump. Of course, all
this could change; we should remember that in 1919 there was only
one gasoline filling station in Britain.

Whether countries decide to scale up battery electric or hydrogen
transportation, they will have to deal with significant infrastructure challenges. 'You cannot see the energy side and the transportation side separately. You have to see it together,' emphasises Zetsche.
Moreover, in order for us to achieve the goal of reducing greenhouse
gas emissions, the power supplied to electric vehicles must come
from low-carbon sources. Electricity distribution infrastructure will
need to be expanded, too. All this means that bold forecasts about
the dominant propulsion systems of the future should be viewed
with some scepticism. 'It would be stupid to say today that the combustion engine is dead,' Zetsche explains, pointing out that progress
towards even more efficient gasoline engines is continuing apace.
There is also the possibility of creating carbon-neutral liquid hydrocarbon fuels, whether by biological or chemical means. 'None of this
should be dealt with in an ideological way,' he advises prudently. He
warns governments against using their regulatory powers to impose
strict controls on gasoline engines to encourage any single solution.
'It would limit all the creativity,' he says. Instead, Zetsche believes
that engineers will come up with a rich array of choices that fit local
needs and personal choices. Giving an example, he feels that plug-in

hybrids with a battery range of 100 or 150 kilometres and a gasoline engine that kicks in only beyond that point may be the most efficient choice for many people, because it would mean that 'for 95% of your journeys, you don't carry the weight of the part of battery you don't need'.

SATELLITES SHOW US THE WAY

Long ago, and before I was able to drive, I embarked on many car journeys with my parents, equipped with maps and guidebooks to areas we would pass through. As the navigator, I felt obliged to plan each journey carefully, but graduated as quickly as I could from this role to that of driver and bought a small Fiat 500 – navigating, a role

FIGURE 7.8 Get lost, without these invisible guides: GPS satellites circle 20,000 kilometres above us.

that has since been taken by machines, held little joy and a lot of potential for criticism and time wasting.

Essential to this new age of assisted navigation are the constellations of global positioning system (GPS) satellites, which orbit 20,000 kilometres above the Earth in the Van Allen radiation belt.[37] On the phone from California, Brad Parkinson, Professor Emeritus at Stanford University, tells me how he conceived the first GPS system back in 1972. After failing to gain approval for a design he had inherited, Parkinson assembled a small team of engineers and arranged a redesign meeting in the Pentagon over the Labor Day weekend, which, he says, 'turns out to be a very quiet time. No phones are ringing, nothing is happening.' By the end of the weekend they had laid down the basic design that underpins all four of today's functional GPS-like constellations.

The original GPS network is a prime example of a 'dual use' technology, one that has both military and civilian uses (an important concept that we will revisit in Chapter 8). 'Our first goal was to allow pilots to drop five bombs in the same hole,' says Parkinson. This military impetus was critical for turning such an ambitious scheme into useful reality. I was very surprised to hear him say that 'even large companies can't afford to make investments that don't bear fruit within a year', but when he explained that it had taken over twenty years to make the system fully operational, I understood. From the start, his secondary goal was to build an inexpensive device that anyone could afford. However, the first GPS receiver that his team built in 1974 cost around $750,000. His ambition was to bring that figure down to $10,000 – it's incredible that a much more capable receiver today costs less than $2.

From the start, Parkinson had a clear sense that an accurate satellite-based navigation system could change the world, but even he is surprised by just how ubiquitous and crucial applications of GPS have become. The system allows anyone to navigate an unfamiliar city with confidence. It helps track freight, guide aircraft, tend crops, coordinate rescue efforts and much more besides, and it can also tell the time: 'The four-satellite technique [upon which GPS relies] gives nanosecond worldwide-synchronised time at all the receivers,' Parkinson explains. The ability to synchronise time became an unexpectedly important attribute, essential for the Internet, cell phones, high-speed trading and the operation of the electricity grid. The world we live in today is completely dependent upon GPS; it is, as Parkinson frames it, 'infrastructure's infrastructure', always available to be used. However, Parkinson is concerned that too many people take that for granted. A keen sailor, he still manually checks his boat's position using dead reckoning, piloting and depth measurements, in spite of having GPS-enabled devices. 'We're losing our skill at map reading. We're losing our skill at getting around,' he says, 'and it worries me a great deal. Having a backup, or contingency, or alternative is very important.'

A similar train of thought caused the European Union, China and Russia to launch their own GPS constellations. As Paul Verhoef, Director of the European Space Agency's Galileo GPS programme, tells me, 'We can no longer allow any downtime on this, because the economic dependence on it is so enormous.' He says that the global market for satellite navigation products and services is predicted to grow to €220 billion annually by 2020. In this light, the €700 million annual cost of Galileo, which is shared between the European Union member states, appears to be a very good investment. New systems like Galileo do more than give us reliability; pooling signals from different systems allows us to measure position with quite astonishing accuracy. Parkinson explains how it is now possible to measure the position of the Earth's tectonic plates with sub-millimetre precision in three dimensions, which could eventually allow for earlier warnings of impending earthquakes. 'As long as [the tectonic plates] are moving, you're fine. It's when they stop that you get earthquakes,' says Parkinson. Even more significant to most people in their daily activities, the steadily improving accuracy and reliability of GPS systems allows a fast-moving vehicle to pinpoint its location to within less than fifty centimetres. As Verhoef notes, this 'not only tells you on which side of the road you're driving, but also in which lane you're driving'. In the accelerating race to develop autonomous automobiles, this ability to locate a vehicle in space will be critical.

THE GREAT CHALLENGE OF THE AUTONOMOUS AUTOMOBILE

'Frankly, if I knew there was a driverless car on the road near me, I would take another route,' says Professor Missy Cummings of Duke University, North Carolina. She is no Luddite but is, in fact, one of the leading experts on control systems for autonomous vehicles, having learned about the shortcomings of these systems through painful first-hand experience. Cummings was one of the US military's first women fighter pilots and in 1989 successfully landed a jet fighter on an aircraft carrier for the first time, yet her exhilaration was short-lived – the next aircraft to attempt the landing was piloted by a friend, who crashed and was killed instantly. 'Pretty much the entire time I flew, I lost one friend, or at least somebody I knew, every month,' she says. She is certain that many of those deaths were caused by the inadequate performance of complex systems that had been designed 'to make it easier to be a pilot, or certainly safer'.

For more than a decade, Cummings flew the world's fastest and most technologically advanced aircraft, including the F/A-18 Hornet. Since leaving the armed forces in 1999, she has been committed to designing and building ways to improve the inter-action between humans and robotic systems. 'My goal is to help usher in this new age of autonomous systems, safely and product-ively,' she says. Much of her attention is focused on the design of autonomous automobiles. 'We would stand to benefit tremendously if we took your average human out from behind the wheel,' says Cummings. In addition to the potential that autonomous vehicles possess to save lives, she believes that they could eventually ease con-gestion and reduce time wasted in traffic jams, but she also thinks it is still too early to release them onto public roads. Concerned that 'the business case is the tail wagging the dog', she feels that this new technology is being pushed into use before it is sufficiently proven to work.

An autonomous vehicle needs to master two skills. Firstly, it needs to perceive and understand the environment around it. Secondly, it needs to be able to predict the near future. Humans do both these things effortlessly, but they are huge challenges for machines. When it comes to perceiving the environment, a key obstacle is the 'brittleness' of computer vision – as Cummings explains, 'A stop sign with a half-inch of snow on it does not look like a stop sign to a self-driving car.' Recently, researchers in computer vision experimented by adding a few small pieces of black-and-white tape to the front of a stop sign; the camera vision system of an autonomous vehicle misidentified this as a sign to keep its speed below forty-five miles per hour.[38] Paul Newman, a founder of Oxbotica, a company that provides autonomous vehicle software, acknowledges that false positives can create problems for these perceptual systems. He describes how a vision system in a car can mistake a high-fidelity photograph of a person for an actual person – making the case for the use of multiple sensor types and a deeper understanding of the scene. Amnon Shashua, founder and CEO of Mobileye, one of the leading manufacturers of autonomous driving systems, points out that false positives of this kind are 'a catastrophe for carmakers'. He goes on to explain that, if you are driving normally along a road and your car unexpectedly engages its automatic emergency braking system,[39] 'you will be scared. The car behaved in an unpredictable way. You'll go and give the car back to the dealer and tell them, "I don't want anything to do with this car model."'

Attempting to predict the future brings another substantial set of problems; Cummings explains that the machine learning algorithms used in autonomous vehicles are intrinsically probabilistic, which means that, when approaching an intersection that they have visited before, under identical conditions, they will not always make the same decision. This measure of unpredictability makes it difficult to design rigorous safety testing regimes. The algorithms that drive these automobiles are usually 'black boxes' and, as a result, the reasoning behind the decision-making process cannot be traced. One specific type of decision that these autonomous vehicles may eventually have to make involves moral dilemmas like the 'trolley problem' proposed by the philosopher Philippa Foot in 1967: when confronted by a child who unexpectedly falls onto the road, should a trolley car – or in this case an autonomous vehicle – risk the life of its occupants and other road users by swerving into oncoming traffic, or should it brake hard but accept that the death of a small child is inevitable?

Cummings has little time for these debates, which she refers to as 'armchair philosophy run amok'. She thinks the technology is still too primitive to be capable of making such sophisticated trade-offs. 'We are not even close to an automobile reliably understanding that it sees a child, versus a bicyclist, versus another automobile, versus a wall,' she explains. Newman's view is that, while it is important to consider how machines should best make decisions in no-win situations, the trolley problem has disproportionate importance in the debate; after all, at the moment of an accident, how many humans remember who they decided to collide with and why? That is not to say that we should not think deeply about the trolley problem, but that we should do so while remembering that the vast majority of accidents are caused by human failure.

Cummings believes that the more pressing ethical debate about autonomous vehicles concerns whether people can be participants in an experiment without their informed consent. As she explains, many cities in the US and around the world have recently approved the testing of autonomous vehicles on public roads and 'without the babysitter in the seat'. She regards these tests as potentially lethal as well as unethical, given that they involve all the other users of a road, none of whom have given their informed consent to being part of an experiment. As I write this, autonomous vehicles have already been involved in lethal accidents on public roads. Zoubin Ghahramani, Chief Scientist at Uber, acknowledges the practical challenges of creating safe autonomous systems but is optimistic about their chances of success. He notes that 'we don't have eyes in the back of our heads' but that autonomous vehicles do, since they can be equipped with sensors, cameras, radar and LiDAR, a system that uses the reflections from pulses of lasers to build up three-dimensional pictures of the environment. Autonomous vehicles also have far better reaction times than people, and they never get tired. When autonomous automobiles are able to communicate both with one another and with the traffic infrastructure around them, they will be able to know what lies around the next corner – how many vehicles are approaching, where they are going and how fast they are travelling. Shashua is developing a system that builds on this; he plans to pool the information gathered by sensors from automobiles made by several different manufacturers and use it to create highly detailed maps of the road. By crowdsourcing this process, it should be possible to deliver maps that update themselves immediately when roads change, due to new signs or construction work – a crucial advance, because autonomous vehicles will rely heavily on this data.

Paul Newman shares Ghahramani's optimism and believes that the arrival of autonomous vehicles is simply a matter of time, because 'we know about sensors and maps and safety. I don't see any glaring hole.' He points out that they will be used first in controlled environments, where potential hazards are easier to predict and accommodate; their success will determine how soon they will be able to work in complex environments like city centres. Many airports, university campuses and commercial ports already make use of them. Durrant-Whyte describes a mine in Pilbara, Australia, which uses 150 autonomous trucks, freeing people from dangerous, dirty and repetitive work. 'Now you can run all the mines across the world from one room,' he says.

As we saw in Chapter 3, people do not trust 'black boxes'. Andrew Adonis points out that people are generally more accepting of accidents that are caused by human error than those caused by errors in engineered systems. 'We already see this in the contrast between public acceptability of road accidents and rail accidents,' he says. 'It's only going to be acceptable to have autonomous vehicles in towns and cities when there's a virtually zero accident rate.' Newman acknowledges that the elimination of accidents will be very difficult to achieve, but hopes that society can engage in better-informed and more level-headed debate about this topic. 'Could we for once, please, be on the front foot about the effects of technology? When we react to genetic modified foods or the risks associated with nuclear power, we are on the back foot and the argument is already lost.' He believes that it is more responsible to 'expect to reduce accidental injury on the road by 80 per cent over the next twenty years, but acknowledge that there will be accidents on the way as the technology matures'. The attribution of responsibility when something goes wrong is also a critical issue, not least for insurance purposes. As Newman points out, 'To say "I showed a black box some data and it learned from it and then did what it did" is not a very good reason. The black box as it exists now isn't really viable. We must develop devices and software stacks that explain what they did and why.'

Shashua is scornful of today's dominant method of testing, which is to get autonomous automobiles on the road and hope to amass a large number of incident-free miles of driving, in order to provide a statistical assurance of safety. He points out that this method is doomed: it will take far too long and, in the meantime, will expose all other users of the road to potentially lethal self-driving prototypes. His solution is to develop a logical mathematical framework that defines what situations on the road are dangerous, and ensures that autonomous automobiles will never make decisions that cause those situations to arise.[40] 'Guaranteeing that you'll not be involved in an accident is not possible, but you can guarantee that you will not *cause* an accident. We can then define blame because, if you [as a human driver] did not respond properly, you are to blame in an accident,' he says. Shashua points out that today's statistical method of safety testing also encourages timid, defensive driving, which could be dangerous – for example, if it frustrates other road users. 'Society will not accept a city like Boston being clogged with tens of thousands of "grandmother-like" driving vehicles,' he says, with a laugh. He is confident that his model-based safety will permit more assertive driving. Partly for that reason, his company is conducting most of their on-road testing in Jerusalem; 'The Israeli drivers are relentless,' he explains. 'If you're not assertive, you'd better stay at home. Nobody is going to give you a gap to enter. You have to push your way in.'

There are other issues, too. Truck and delivery vehicle driving is the most common source of employment in twenty-nine out of the USA's fifty states;[41] will autonomous vehicles have a significant impact on unemployment? Cummings believes that much of the concern regarding jobs is 'alarmism' and says, 'These are enabling technologies that will cause growth in a lot of other areas. The one thing that we never can do well is forecast new jobs.' Full autonomy will also affect the place of the automobile in society. At present, the average car is parked and idle 95 per cent of the time,[42] but if a car no longer depends on a human driver it could serve many more people, and shared ownership could become much more common. According to MIT Robotics Professor Daniela Rus, 'If self-driving vehicles penetrate society the way we expect, then we will have mobility on demand – transport as a utility.' However, as appealing as this vision is, it can only be realised if automotive engineers, businesses and regulators engage deeply with the hopes, fears and foibles of the people who will be using these robotic vehicles.

THE TRANSPORT OF THE FUTURE

The flying car has for decades been a common image in science fiction. In his satirical vision of the future of transport in 1829, William Heath even showed a person perched on a small, rotor-powered craft that bears an uncanny resemblance to a modern-day drone. Cummings thinks that, compared to self-driving automobiles, it is probably easier to make autonomous drones and small aircraft that fly safely, partly because it is easier for vehicles to avoid each other in a three-dimensional space. As she says, 'We have been doing automated landing in planes for over thirty years, so there is no question in my mind that the drone is a technology that is here to stay.' Already, drones have been used for many things: in agriculture, they monitor crops and can deliver water, pesticides and fertilisers with great precision;[43] for disaster relief, they deliver vital medical supplies; and in warzones they are used to carry weapons. Cameras mounted on drones have given news reporters, film-makers, surveyors and

FIGURE 7.9 Medicine without borders: a drone flies medicine to a remote Japanese island.

emergency services important new capabilities, though they do further reduce privacy.

Cummings accepts that the list of tasks drones are used to fulfil will continue to grow, but maintains that flying cars are still a long way off: 'The technology is not the problem … the real problem is the infrastructure that goes around it.' The air traffic control systems required to monitor large numbers of passenger drones and light aircraft simply do not exist. Ann Dowling is also sceptical about the idea of flying cars: 'Think beyond the current congestion that we have on the roads; extrapolate that into the skies, with people all being poor flyers, rather than poor drivers. It would be a complete nightmare.' Wind is another serious problem for any transport system that relies on small flying vehicles. 'Having flown light

FIGURE 7.10 Fantastical fantasy: men on the moon, imagined 250 years ago. Engraving by Filippo Morghen (1768).

aircraft, there are lots of conditions in which they just don't work,' concludes Dowling.

Sometimes science fiction forecasts the future with some degree of accuracy. In 1638 John Wilkins, the first person to write about binary codes (which I discussed in Chapter 3), published a text called *The Discovery of a World in the Moone*. In it, he imagines travelling to the moon, where he discovers a population of American Indian people who live in pumpkins and travel on fantastical, flying chariots that are powered by large birds. It is pure fantasy, and to our modern eye his visions are completely bizarre, since they are so clearly informed by Wilkins' seventeenth-century experience of the world. However, although the technical details are wrong, Wilkins was right to predict that curiosity and technological progress would eventually carry us to the moon.

Attempting to predict future technologies is challenging, even for those tasked with building them. Wilbur Wright was wise when he said, 'In 1901, I said to my brother Orville that man would not fly for fifty years. Ever since, I have distrusted myself and avoided all predictions.' Forecasting has always been a perilous business, and the demise of Concorde is revealing in this respect – when it first flew, many saw it as a sign of things to come, believing that air travel would get progressively faster. Dowling thinks that supersonic passenger planes may well be back in the air at some point soon, but never as mass transportation for people or freight. For large aircraft, 'it's too hard to dissipate the sonic boom', Dowling explains. She does, however, believe that supersonic flight might be possible for smaller business jets, and is also open to the eventual possibility of 'hypersonic flight', whereby aircraft would travel well above the stratosphere at speeds of at least Mach 5, reducing flight times significantly. Ralph Robins describes a new generation of hybrid rocket and jet engines under development that can use atmospheric oxygen up to about 80,000 feet, before switching onto stored oxygen. However, to be practical, these engines would have to achieve a massive, and near-instantaneous reduction in temperature – as Robins remarks, 'The heat exchanger alone would probably cost more than the airplane and everything else.' But these vehicles, just like rockets that could one day take passengers to Mars and back safely, remain the stuff of exciting science fiction.

Significant advances that occur rarely, such as the first plane flight or the first manned rocket, always capture the public's imagination. While engineers attempt to make possible the next great leap into the future, many others will have more impact by making less dramatic, but equally important, incremental innovations. Think of the stirrup, the improvement of asphalt road surfaces or the way in which automobiles gradually became safer and more efficient. Dowling's view is that 'we should really put more emphasis on the door-to-door time, or efficiency of a journey, rather than just how fast you go through the air'. In other words, the key challenge, rather than making vehicles that travel faster, should be building integrated transport systems that make our movements around the planet more efficient, cheaper and less frustrating. Dowling explains that we constrain ourselves by funnelling all flights through air lanes, 'which are essentially like motorways laid down in the sky'. The ability to use more of the sky would 'free things up hugely', she adds. She thinks that intelligent systems should be continuously able to track where aircraft are, which would allow flights to take more direct routes and eliminate holding patterns that waste fuel and cause noise pollution around airports. 'We're still using very old systems and we're not using the flexibility of the technology that's here,' she says.

These are the sorts of challenges that transport engineers have been grappling with for centuries, as they steadily improve the transit of goods and people. Epidemiologist David Bradley describes the progressive democratisation of travel in an ingenious way.[44] Looking back at his family history, he writes that his great-grandfather rarely left the English town of Kettering and never ventured more than twenty-five miles from home. His grandfather, with the help of the rail network, routinely travelled to London and other cities around the UK. His father could reach out further still, visiting much of Europe. Bradley himself, with thanks to jet aircraft, regularly travels across the world for work and leisure. In each new generation, the average person has been able to travel ten times further than his or her parents.

In the last chapter I talked about how trade began with agricultural surpluses. Trade needed transportation, which led to the mixing of people and ideas. It allowed us to satisfy our human urge to explore not just Earth but also the moon, as well as the planets beyond. The way we travel has become more efficient and more convenient, as progress took us from the horse to the railroad and automobile and eventually to the airplane and the rocket. Over time, most vehicles will become autonomous, a great advance that will improve the safety of transportation. Furthermore, I expect that fossil fuels will soon only power airplanes and the like; the rest will be increasingly powered by electricity, as batteries become more efficient and as we are able to generate enough clean electricity – this transition will represent a positive contribution to a cleaner world with a stable climate.

FIGURE 7.11 Beam me up, and beam me down: teletransportation made easy for Captain Kirk and co., *Star Trek* (1966).

Efficient and convenient transportation does not make the world smaller, but it does make it bigger, richer and more replete with possibilities for exchange and growth. When I first visited Beijing in 1979, I arrived in a country that seemed like it had been closed to the outside world for generations. On that visit I felt a bit like Marco Polo, the first European to visit China, as the local people, all wearing grey-green Chairman Mao suits, stopped to stare at me. These days, things have changed dramatically; the house in Venice that is reputed to be the home of Marco Polo, not too far from my own home, is usually crowded with Chinese tourists, and Beijing is a thriving cosmopolitan city. No Chinese national finds the presence of Europeans unusual. Even with the limitations of our technology, which will never match the fantasy of *Star Trek*,[45] our ability to move around the world has changed everything, and innovations in this area are sure to revolutionise our future.

8

Defend

The roar of the Vulcan bombers' jet engines seemed to shake the ground as they streaked across the sky above me; the exhilarating noise and the trails of red, white and blue exhaust left by the arrowhead-shaped aeroplanes are among the most vivid of my early memories. I was a young boy watching the celebrations for the coronation of Queen Elizabeth II in Singapore in June 1953 – I was there because my father, an army officer, was stationed at the UK military base. It was the time of trouble in what was then called Malaya, as the Malay, rebelling against British rule, fought for the independence of their part of the peninsula. My father's predecessor had been killed by terrorists, so we lived with very tight security in a big house on Tanglin Hill, a leafy district of Singapore. My other lasting memory is of being taken to my primary school in an armoured car – war and the military were ever-present in my childhood.

As I watched the fly-past on that humid day, my father was on the parade ground with his soldiers. As a five-year-old, I was oblivious to the abysmal reality of war; the planes that roared overhead were a straightforwardly exciting spectacle, and I had no any inkling that they had been designed to carry the most devastating cargo of all. Just eight months before the coronation festivities, the UK had become the third country, after the US and the USSR, to start building an arsenal of nuclear weapons; the Vulcan bombers represented a core part of its nuclear deterrent.

Ever since a stone hand axe was first used in anger, people have had weapons. The need to defend that which is yours, and the temptation to gain through violence that which is not, have been prominent concerns throughout history. This fundamental characteristic has not only caused bloodshed and devastation – it has also, perhaps counter-intuitively, delivered real progress. The statistics show that most people, in most parts of the world, are now less likely to be involved in war than ever before. Beyond helping to make our world safer through stronger defence and better deterrence, warfare has always increased the pace of innovation. As a result, many of our most potent technologies, be they in medicine, communication, transport or elsewhere, have their roots in military engineering. Without the immense investment in government-led defence, we would not have jet travel, antibiotics, the Internet or many other marvels of the modern world. When considering engineering for defence, simple-minded notions of 'good' and 'bad' do not suffice. And, as humankind's destructive capacity has grown, so too has its ability to create.

The hideous power unleashed by the atomic bombs that devastated the Japanese cities of Hiroshima and Nagasaki in 1945, killing at least 130,000 people, demonstrated that destructive capacity approaching its apex. I only appreciated the full horror of this when, as an adult, I visited the Peace Memorial Park in Hiroshima and saw a burial mound containing more than 70,000 porcelain canisters, each one holding the cremated remains of a victim whose body parts could not be identified. Most moving, for me, were the accounts of the explosion's aftermath, either written or drawn by survivors. The deeply personal ways in which these most vivid and disturbing of memories were communicated rendered them more powerful than any photograph or video. Confronting these raw and unfiltered accounts, which included images of bodies with skin hanging off in strips, others charred beyond recognition and piles of children's corpses, was profoundly moving and made it absolutely clear to me that nuclear weapons must never be used again.

FIGURE 8.1 Never forget, never again: a survivor's image of victims of the atomic bomb on Hiroshima (August 1945).

THE MOST SPECTACULAR EVENT THAT NEVER HAPPENED

My generation was the first to grow up with the knowledge that engineers had created objects that could, at the push of a button, kill everyone on the planet. Ever since then, humankind has dangled precariously over an abyss; a single false alarm, a technological glitch in a control system or the election of a reckless leader to a nuclear state could provoke a war that would end all progress. There have been plenty of near misses, but somehow the most hideous event imaginable has been avoided, which is what the US Nobel laureate and economist Thomas Schelling meant when he said that 'the most spectacular event of the past half-century is one that did not occur'. Schelling's contemporary, the mathematician John von Neumann, was a hawk who, soon after the Second World War, advised US President Truman to launch a huge nuclear attack on the USSR before they could acquire atomic and thermonuclear weapon capabilities themselves.[1] According to von Neumann's logic, if the US waited, nuclear war between the world's new superpowers would be inevitable. Schelling instead pointed out that these weapons could act as a powerful deterrent, since they meant that any declaration of war between the two states would be an act of joint suicide. Partly because of Schelling's intervention, no nuclear weapons have been detonated in anger since those that ended the Second World War. Amid the tensions of the Cold War that followed, the number of nuclear warheads rose rapidly, but those stockpiles are now shrinking. While 70,000 warheads existed in 1986, today there are fewer than 15,000[2] – enough to cause catastrophic damage to our world, but the declining trend, as well as the fact that nobody has turned the nuclear launch key for seven decades, are reasons for hope.

Surprising as it may seem, we are fortunate to live in unusually peaceful times. While the news brings fresh reports of conflict in Syria and Yemen, these are increasingly unusual events, but that was not always the case. Prehistoric human bones show that in some communities as many as one in six people died violent deaths, mostly at the hands of other humans.[3] By the late Middle Ages, however, the rate of violent death had decreased to an average of one in 2,000 people. In the twentieth century there have been two world wars and several acts of large-scale murder – in Hitler's concentration camps, in Stalin's gulags and during Mao's Cultural Revolution. The Great Famine resulting from Mao's policies killed more people than the First World War. These events stand out as dreadful exceptions to a trend of declining violence – while horrifying conflicts still erupt in parts of our world, the statistics clearly show that during every decade of my lifetime, wars between states have become less frequent, with each war claiming fewer lives.[4] Over the last three decades, wars of all types have resulted in an average of two deaths per year for every 100,000 people alive,[5] a small fraction of the number who die, for example, from complications associated with obesity.[6]

This is all encouraging news, but we cannot afford to be complacent. While the doctrine of mutually assured destruction has helped to maintain a fragile peace, the threat of nuclear Armageddon has

RATE OF VIOLENT DEATHS IN WAR

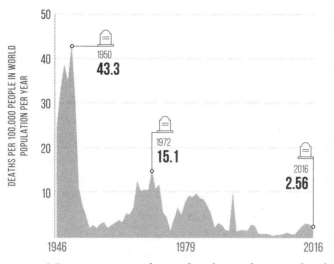

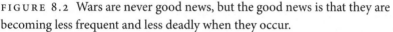

FIGURE 8.2 Wars are never good news, but the good news is that they are becoming less frequent and less deadly when they occur.

not completely gone away. If defence in the twentieth century was characterised by symmetric conflicts and stand-offs, countries must now repel more varied attacks, by less equally matched opponents. As Hugh Durrant-Whyte, Chief Scientific Advisor at the UK Ministry of Defence, explains, the boundary between war and peace is becoming increasingly blurred: 'People don't declare wars anymore ... things are more nuanced.' States are not alone in having destructive capabilities, and terrorists and other non-state actors have many methods to kill and disrupt: explosives, vehicles driven at people, cyberattacks on infrastructure, deadly pathogens and autonomous weapons with specific targets. So, while progress towards a peaceful world continues on many fronts, it is far from complete.

THE CIVILISING INFLUENCE OF GUNPOWDER

In Europe, the first decisive shift towards greater peace began with the introduction of a new and powerful weapon that the pre-eminent British Sinologist Joseph Needham called 'one of the greatest achievements of the medieval Chinese world'. Gunpowder was invented by Taoist alchemists towards the end of the ninth century, as they searched for substances that could extend human life. They warned, however, that mixing the substance's ingredients – saltpetre, sulphur and charcoal – had to be done with caution because 'some who have done so have had the mixture deflagrate, singe their beards and burn down the building in which they were working'.[7] The military potential of gunpowder became obvious during the tenth century, when gunpowder weapons became a standard feature of Chinese warfare – some 300 years before they appeared in Europe. Over time, a sophisticated and deadly array of bombs, 'fire lances', rockets and primitive cannons were invented. Professor Jianjun Mei, Director of the Needham Research Institute in Cambridge, explains that historians cannot confidently reconstruct the routes by which gunpowder technology spread out from China, but thinks it is most likely that it first reached the Mongol Empire and then spread out through great swathes of Eurasia.

Once gunpowder arrived in Europe in the thirteenth century, it found immediate uses. In 1267, the philosopher Roger Bacon described how, when gunpowder explodes, 'so horrible a sound is made ... that we find the ear assaulted by a noise exceeding the roar of strong thunder, and a flash brighter than the most brilliant lightning'.[8] It was clearly a frightening substance. During the fourteenth and fifteenth centuries, engineers learned to build increasingly accurate and powerful cannons and muskets, and these new weapons irreversibly changed the nature of warfare. Needham reports that in 1449, an artillery train controlled by the French king moved through the province of Normandy, systematically destroying all the remaining English castles. In order to defend cities, fortifications had to be built more robustly – if they could withstand artillery, they could also keep infantry attackers at bay, and another consequence was that conflict could be waged at a greater distance. These technological advances meant that war became more expensive and could no longer be waged by individual nobles but only by kingdoms and nation states. That heralded the beginning of the end of the military feudal system

FIGURE 8.3 Ancient boom! Demons playing with fire and gunpowder in a Chinese silk painting (AD 950).

that had held sway throughout much of Europe during the Middle Ages, as more centralised and organised forms of power gained the upper hand. The emergence of stronger governance encouraged the development of enforceable systems of law and justice, which reduced violence and enabled knowledge and wealth to grow more securely. The nineteenth-century historian Henry Thomas Buckle went so far as to credit gunpowder with triggering the emergence of the 'intellectual middle classes' – since gunpowder weapons were difficult to handle, they led to the professionalisation of warfare, which, he argued, meant that a smaller proportion of the population were occupied with fighting and the rest could therefore devote their intellect and energy for more constructive endeavours.

SHARING THE FRUITS OF WAR

Engineering for defence has always done more than just keep us safe from attack; war also encourages engineers to expand the boundaries of the possible, and when survival is at stake, the relaxation of cost constraints can also allow more creative developments. It is striking just how many innovations have had their roots in military engineering. From the stirrup (the impact of which I explored in the last chapter) to the Internet, defence has delivered many transformative technologies. More recently, the fundamental engineering behind the touch screen, the voice-activated assistant and many other core components of the smartphone all began life within the defence sector.[9] Nigel Whitehead, Chief Technology Officer at BAE Systems, says, 'Almost every major technological advance that's ever been made has been justified on the basis of achieving military advantage.' He believes that engineering in the service of defence continues to steer the course of innovation, an idea 'that either sits comfortably with you or it doesn't'. Whitehead reinforces his point that almost every military technology ends up being used for civilian purposes; this concept of 'dual use' goes back a long time.

In the third century BC, Archimedes used his discoveries in mechanics and hydraulics to build irrigation systems and ships, as well as a range of weapons. In the fifteenth and sixteenth centuries, Leonardo da Vinci created some of our most beautiful and humane works of art, but he also channelled his unique intellect into conceiving engines of war,[10] blurring the boundary between artist and engineer, and between the aesthetic, the practical and the destructive. The Venetian Republic invited him to Venice, to advise them on ways to defend their city against attacks from the Ottoman Empire. He understood that geography often provides the most powerful defence, which is why Venice had been founded on a series of islands in a shallow lagoon; even today ships often run aground on the sand bars that surround the city. Da Vinci advised the Venetians to lure the Ottoman forces into the Isonzo river valley, sixty kilometres to the east of the city, and then to remove a series of temporary dams, flooding the valley. He also drew plans for a type of submarine and a diving suit, which he suggested the navy could use to approach Ottoman ships by stealth, before boring holes into their hulls to sink them.[11] Venice did not implement da Vinci's plans, although a later reconstruction showed that his diving suit would have worked.

The Arsenale, a naval shipyard, was established in Venice during the twelfth century and by da Vinci's time it occupied nearly a fifth of the city's land and employed 16,000 people, roughly a sixth of the population. The elegant brick-built entrance, captured so evocatively in Reuwich's fifteenth-century woodcut (see Figure 7.2), is one of the city's first and best examples of classical revival architecture. Today, most people know the Arsenale as one of the venues for the art Biennale, but for me it remains most significant as a symbol of Venice's past commercial and military power – the ships produced there were the primary drivers of the city's rise to prominence as an influential centre for trade.

In the eighteenth century, the economist Adam Smith described the benefits of specialisation and division of labour, methods that had been put into practice centuries earlier by the naval engineers of the city's Arsenale. At its peak in the early sixteenth century, the shipyard could construct a complete galley every day. Activity was organised so that a partially assembled ship moved from a dry dock into a canal dock, before being floated to different stations, where teams of workers executed specific manufacturing steps. Shipbuilding there was a triumph of efficiency that would not be equalled until the nineteenth century; the large naval fleets launched by medieval Venice allowed it to control much of the Mediterranean Sea. The Republic stood at the crossroads between East and West, and soon came to control the import of exotic luxuries such as spices, sugar, silk, lacquer and precious stones into Europe, and the export of German minerals and British woollen goods to the East. From the beginning, the Venetian galley was effectively a marriage between a fighting ship and a trustworthy merchant vessel; it was itself a 'dual use' technology.

Kevin Tebbit, a former Permanent Secretary of the UK Ministry of Defence, is clear that it is often impossible to disentangle the motivations behind the great innovations – he wonders 'does trade follow the flag, or does the flag follow trade?' Tebbit sees the Arsenale as a case in point, where the military and civilian use of a technology advanced hand in hand. Without the ability to defend their valuable cargo from pirates and rival fleets, Venetian traders could not have generated the wealth that allowed their Republic to flourish.

Venice's shipbuilders also continued to innovate so that their navy could maintain its supremacy. They were the first to equip their galleys with cannons, at the Battle of Chioggia in 1380, but it became clear that these ships, which were mainly powered by teams of rowers, could not carry enough firepower. As a result, cannons started to fill more of the space previously occupied by oarsmen, who were increasingly replaced by sails. These changes led to a more advanced vessel called the galleon, introduced during the sixteenth century. These multi-decked sailing ships sported multiple masts, ranks of cannons and raised defensive castles at the front and rear. The design proved so successful that it was copied by other navies; these ships would dominate the high seas until the nineteenth century.

Galleons launched a new era of exploration and international trade, and it was these ships that returned from the expeditions to Latin America laden with gold and silver. They were also used in the triangular route of the transatlantic slave trade. Ships went from England to West Africa with textiles, alcohol and guns to exchange for slaves. Then they sailed across the Atlantic to the New World plantations, where the slaves were exchanged for sugar, cotton, coffee and tobacco, goods that were brought back to England and sold, generating great wealth for merchants, at huge humanitarian cost to the slaves and their families.

As more of the world was discovered, maritime charts became more precise, the compass – another ancient invention with Chinese roots – became more accurate and the mathematics needed for dead reckoning was established.[12] The latitude of a ship could also be determined by measuring the elevation of the sun or certain stars above the horizon. However, despite these advances, long-distance oceanic travel remained a perilous endeavour, and one of the principal reasons for this was that the accurate determination of longitude remained impossible. In 1707 twenty-two British naval ships were

FIGURE 8.4 Anything but loose cannons: Venice's Arsenale built the first ships to carry full batteries of cannons. Engraving by Coronelli (1693).

wrecked on the Isles of Scilly when their navigators were unable to calculate their longitude in the dark and stormy conditions, resulting in the death of 1,500 sailors. This limitation threatened the safety and effectiveness of the British Royal Navy and so, under the terms of the 1714 Longitude Act, it decided to fund a prize of £20,000 (or over US $4.5 million in today's money), for anyone who could invent a way to determine longitude with an accuracy of no more than thirty nautical miles.

Framing a challenge as a competition can encourage innovators to break out of previous mental assumptions. When it came to determining longitude, accepted wisdom dictated that the solution would most likely involve an astronomical device. The eventual winner of the competition, however, realised that the most straightforward solution was, in fact, to build an accurate clock that was reliable enough to hold its time during long voyages at sea – for if a ship can compare its local time with that in its place of departure, it is easy to convert that time difference into the true distance covered east or west, that is, its longitude.[13] Calculating local time was straightforward – sailors had been doing that with the astrolabe since antiquity – but the problem was keeping track of the time at the port of origin. Even Isaac Newton doubted that a sufficiently accurate clock could ever be built, because 'by reason of the motion of the Ship, the Variation of Heat and Cold, Wet and Dry, and the Difference of Gravity in different Latitudes, such a watch hath not yet been made'.[14] However, a carpenter-turned-clockmaker named John Harrison was able to craft a winning mechanism.

I went to see Harrison's H4 chronometer at the National Maritime Museum in London. This large watch, which measures about five inches in diameter, is housed in a beautiful silver case, but it was the nature of the mechanism inside that prompted Harrison to boast 'that there is neither any other mechanical or mathematical thing in the world that is more beautiful or curious in texture than this my watch or Timekeeper for the Longitude'. Harrison did away with a pendulum, which could not be stable when on a ship, and instead invented a mechanism that minimised friction and did not need to be oiled. An unusually large balance wheel oscillated backwards and forwards, faster and more powerfully than earlier watches, helping to keep time and resisting the disturbances created by the pitching ship and the changing gravity and humidity. As a result, the watch ticked at a notably rapid five beats per second. Harrison also introduced a bimetallic strip of steel and brass that helped compensate for changes in temperature. The watchmaker used 'devices which no clockmaker has ever thought of using' and succeeded by 'tackling his mechanical problems as an engineer might, and not as a clockmaker would'.[15] It took several decades for him to perfect his marine chronometer, but the Navy's prize had achieved its aim. Despite the H4 being judged a success on a naval voyage from Portsmouth to Jamaica in 1761, Harrison had to fight long and hard for the prize money, which he only received a dozen years later. Military imperative drove the solution to the longitude problem; in Tebbit's view, Harrison's invention of a new and reliable way of keeping time is one of the clearest examples of 'a

FIGURE 8.5 Calculating longitude was no longer a wild guess with John Harrison's H4 marine timekeeper (1759).

precise point in time where a defence-funded innovation changed the world profoundly'. Accurate timekeeping opened up the world to trade and exchange, and the order and efficiency delivered by the clock helped the Industrial Revolution gather momentum, allowing raw materials and workers to come and go on a predictable schedule. Manufacturing procedures advanced rapidly, changing every aspect of life, not least the course of warfare.

THE RISE OF INDUSTRIAL-SCALE WARFARE

By the end of the First World War, the technological landscape had profoundly changed. The cavalry that had been driven helplessly into machine gun fire at the start of the war was replaced by a mechanical innovation called the tank. A new form of killing machine, it was designed to combine heavy firepower with near-impenetrable armour and unstoppable mobility. The first models were mechanically erratic, but improved considerably as their internal combustion engines became more efficient and reliable. Those same engines made trucks more reliable, too. In addition, the advances in chemical engineering that produced poisonous gases and potent explosives also led to the production of ammonia-based fertilisers, which subsequently saved millions from starvation. The urgent need to coordinate military actions and gather intelligence stimulated great improvements in the reliability of telephones, radio communications and wireless telegraphy. The legions of wounded and maimed survivors provided great scope for medical advance, with army medics such as Surgeon Captain Harold Gillies pioneering new ways to patch up and reconstruct skin, bone and connective tissue, creating the new field of plastic surgery.

During the Second World War, there were even more significant developments; the technological prowess of the military forces on all sides meant that the threat of destruction was greater than ever, but so too was the potential to counter it. The armaments, manoeuvrability and armour plating of tanks was greatly improved. The formidable technological challenge of using radio waves to create radar

FIGURE 8.6 The indestructible killing machine of the time: a British First World War tank.

systems that could forewarn of incoming enemy aircraft was solved by both British and German engineers. The 'Dowding system' used a dedicated network of telephone lines, integrating the huge amounts of information from radar batteries and observation posts all over the country before coordinating a response. In Churchill's view, 'all the ascendancy of the Hurricanes and Spitfires would have been fruitless but for this system ... a most elaborate instrument of war the like of which existed nowhere in the world'.[16] Work on the Dowding system, much of it led by Nobel Prize-winning physicist Patrick Blackett, contributed to the emergence of a new branch of mathematics called operations research which, as we saw in Chapter 3, is one of the foundations of modern-day machine learning and artificial intelligence.

Part of Germany's response to Allied radar defences was to build the world's first long-range ballistic missile. The *Vergeltungswaffe 2* (or 'Retribution Weapon 2'), usually known as the V-2, was a weapon of terror. Launched into the stratosphere, these rockets descended on their targets without warning, and exploded on impact. Encapsulated in them were the core elements of the systems that would be needed to reach space, which is what the V-2's inventor Werner von Braun maintained that he really wanted to do. However, when Hitler had offered him the resources that he needed to advance rocketry, albeit for destructive purposes, he eagerly agreed.[17] After the first V-2 bomb hit London, von Braun was reported to say that 'the rocket worked perfectly, apart from landing on the wrong planet'.[18]

Most of the V-2 rockets that reached Britain fell on London, where they caused widespread fear and killed nearly 3,000 people, a small number compared to the 20,000 slave labourers killed while making them in an underground factory near the Buchenwald concentration camp. Despite his involvement in this horror, von Braun ended up as an unlikely hero of the Space Age; after the war he was brought to the US, where he played a key role in developing the rockets for the Apollo programme that sent the first people to the moon.[19]

'We couldn't live today without the satellites and the technologies that were put in space,' according to Kevin Tebbit. Transport, banking,

FIGURE 8.7 Destruction from outer space: a V-2 flying bomb (*right*) invented by Werner von Braun (*left, alongside the Apollo 11*).

telecommunications, agriculture and many other civil sectors all rely on satellite technology, which has its roots in the military. Launching the first satellite and putting people on the moon was important for national pride and inspiration, but these were not the only motivations – almost as soon as the Second World War came to an end, the race was on between the US and USSR to be the master of space, 'the final frontier'. And rockets that can reach space and then re-enter the atmosphere can also deliver nuclear weapons silently and almost undetectably.

Satellites also proved to be the ultimate 'eyes in the sky' surveillance mechanism. In 1978, the first global positioning system (GPS) satellites were launched to an altitude of 20,000 kilometres. Long before they were connected to our cars and our smartphones, they were used to guide ballistic missiles and to help troops pinpoint their location in warzones. Military research and investment may no longer lead the way in space technology but, from the perspective of defence, we have unfinished business in space.

I meet engineer Ben Reed in a large, high-ceilinged room at NASA's Goddard Space Flight Center that is hung with black drapes to simulate the darkness of space. Engineers cluster around a satellite that they are building and in the distance is a replica of an asteroid. All of this makes me feel like I have stepped onto the set of a science fiction movie. But the satellites here are real, and Reed hopes that one of them in particular will have a big effect when it is sent into orbit in 2020. Reed explains that 'ninety-nine per cent of satellites launched are what we call "lone wolf" craft', which are sent up to perform a specific task and then, when they run out of fuel or complete their objective, either crash back to earth or move into a 'graveyard orbit'. Reed and his team want to change that – he shows me a robotic satellite that will be able to seek out and dock with existing satellites, to refuel them or upgrade their technology. 'This is unbelievably challenging,' says Reed, 'but that is NASA's role: to redefine the possible so that others may follow'.

If NASA could reuse or repurpose more satellites, space-based technologies could become cheaper and more flexible. However, there is a danger that technology designed to service a spacecraft could be used for more disruptive ends. As satellite engineer Martin Sweeting points out, removing debris or adjusting electronics is just one step away from removing or disabling the space hardware of rival countries or companies. Interference with the GPS satellite network, for example, would cause profound disruption; this is one of the reasons why Russia, the European Union and China have each launched their own GPS-like constellations (as I described in Chapter 7). As we rely increasingly on space-based infrastructure, we must make sure that we can also defend it.

Although the technological advances needed to develop rockets, radar and advanced aircraft during the Second World War were extraordinary, none of them matched the ambition of the Manhattan Project. The Allies were concerned that the Third Reich's scientists, led by Werner Heisenberg, the theoretical physicist who had pioneered the theory of quantum mechanics, were developing nuclear bombs. The foundations of quantum physics had only recently been established; theory had shown that atoms contain vast amounts of energy, an idea confirmed by Albert Einstein's equation $E=mc^2$, the demonstration that a particle's energy is equivalent to its mass, multiplied by the speed of light squared. The speed of light, approximately 300 million metres per second, is a very large number, and when squared it becomes a gigantic one. A mass of just one gram has the equivalent energy contained in an explosion of more than 20,000 tons of TNT, enough to power a large modern city for an hour. The conversion of mass to energy can occur when an isotope of uranium is split by bombarding it with neutrons.[20] During the four years of the Manhattan Project, these theoretical implications of quantum physics were confirmed and reduced to clear and applicable mathematical models. The apparatus required to purify the appropriate isotopes of uranium was built, and a device that could carry and then detonate the nuclear material was constructed. By 1945, theory and practice had come together to build and deliver the bomb. But this era-defining military advance also delivered something useful in peacetime; more than 10 per cent of the world's electricity is now generated by safe, low-carbon nuclear fission reactors. Furthermore, biological science was also transformed by these advances; Crick and Watson were barely interested in DNA until biologists in the US had proved, with the help of radioactivity, that DNA carried life's code.[21] A lack of calculating equipment in the Manhattan Project also encouraged John von Neumann to lay down the architecture for a modern programmable digital computer.[22] When this was combined with Turing's work on the decryption of enemy cyphers during the war, the age of modern computing was born.

THE RACE FOR CYBERSECURITY

One cold day in April 2017, I visit the Washington headquarters of the Defense Advanced Research Projects Agency (DARPA), the direct descendant of the research agency at which work on the earliest version of the Internet began in the 1970s. As the modern world relies ever more on the Internet to support every aspect of our lives, a critically important, but almost completely invisible, new battlefield has opened up. I had come to DARPA to find out what that agency is doing to help defend our critical information and infrastructure in the age of cyberwarfare.

After showing my passport and surrendering my mobile phone, I was finally shown through the security gates and into the DARPA offices. As I got into the lift, the public relations manager who was escorting me to the meeting room said, 'Mike's very modest, but he's probably one of the world's best hackers.' When I ask Mike Walker if it is true that all computer systems are vulnerable to hackers, he insists that this is not the case. As an example, he explains that the processor which controls the payment capability built into the Apple iPhone is not hackable – its code, of which there are some 50,000 lines, has been formally verified, which means that programmers have provided a mathematical proof that the algorithm will behave as intended and that it is completely invulnerable. 'The monster is complexity,' Walker explains. The code for the latest Windows operating system is about one thousand times longer than iPhone's payment code, and much more than one thousand times more complex. 'Even if Moore's Law were to continue ... there's no path to building the computer that can formally verify 50 million lines,' he says. And it is not just sheer complexity that can make software vulnerable. Anders Sandberg of the Future of Humanity Institute at Oxford University points out that a lot of code is too sloppily written. This, in his view, is because we want the power and speed of the computers, without wanting to pay for software that is designed to maximise security – and the end result is that 'we accept a lot of complexity that is badly held together'. The inevitable conclusion is that all but the very simplest of computer systems contain an unknowable number of exploitable weaknesses.

As things stand, the battle that is being waged for information and control is a fundamentally uneven conflict. As Walker characterises it, 'the defence is always public, whilst the attack is private'. The aim for most cyber attackers is never to be seen – they can spend as long as they wish searching for vulnerabilities and rehearsing strategies, 'but then they press "go", and victory happens in seconds'. He compares cyberwarfare to a race, where 'the only people that we see in the news are people that have a breakdown, when the wheel has fallen off the car and they've gone into the side of the fence'. The asymmetry of the situation means that even investing in the very best commercial computer defence software cannot guarantee security: 'Just having a playing field where the more expensive capability can achieve victory would be revolutionary. It's not that the defence isn't paying more. They are,' says Walker. Nigel Whitehead believes that the possibility that 'a lone teenage hacker in a bedroom can bring down a nation state' is real: 'They essentially have weapons of mass destruction at their fingertips, if they know what they are doing'. All they need to do to kill people and spread panic is shut down the power supply to a hospital, or turn off the traffic lights in any big city. And by breaking food supply chains they could cause rioting in the streets. Whitehead believes that, even if they don't appear immediately destructive, these threats warrant very serious attention.

A cyberattack on a nuclear missile command and control system does not bear thinking about, yet it is something that MIT Professor Ernie Moniz has to think about a great deal in his capacity as CEO of the Nuclear Threat Initiative.[23] Moniz recalls an incident in 1983 when radar screens in Moscow showed five US intercontinental ballistic missiles hurtling towards the USSR. According to protocol, the USSR should have immediately launched a retaliatory strike. But, as Moniz recounts, 'One guy down the chain said "No, I'm not going to do that. This can't be right." ' His gut instinct was vindicated, when the apparent attack was proven to be a false alarm caused by Soviet satellites misinterpreting sunlight reflected by the tops of clouds. Moniz is worried that a cyber attacker could create even more convincing hoaxes: 'Now, with all the hacking going on, what are the real signals, what are the false signals?' he asks.

Moniz is deeply concerned about the future risk of cyberwarfare in which, unlike in more traditional forms of war, 'There are no rules of engagement, no red lines that we will respect'. As Walker indicated, unidentifiable individuals or states can strike with little or no fear of retribution. He believes that, as they search for solutions, software engineers will increasingly come to view computer systems as complex ecosystems, and will only 'use as much complexity as is safe for the job'. Where safety or privacy is of the utmost importance, the amount of code will be pared down to the bare minimum, and only used if its invulnerability to outside interference can be proven. As he says, 'When we have something really important to do, like make a payment without losing our credit card number, then we have a very simple computer, whose entire job is to ask the user if the secret [number] can be used, use it once and not do anything else'. These simple, un-hackable systems are becoming capable of being deployed in more sophisticated operations – for instance, DARPA engineers recently succeeded in writing a control system for an unmanned military helicopter which, they believe, cannot be hacked.[24]

The threat of war has always accelerated the rate of innovation. DARPA knows that, and during peacetime they try to recreate that threat by creating competitions. That was the aim of the British Navy when they offered the Longitude Prize in 1714. Three centuries later, it was a DARPA competition that catalysed progress in self-driving car technology; in 2004, the first year of the competition, none of the vehicles completed the 150-mile course, with the best performer grinding to a halt after just seven miles. The following year, with the prize money doubled to $2 million, five teams completed the course.

DARPA's most recent competition was designed to find new and effective autonomous cyber defences, which are based on artificial intelligence and therefore need minimal human oversight. Walker, who was in charge of the competition, describes it as 'about colliding communities ... These collisions always pay off ... they force the researchers to do the engineering and to think about application'. Competing teams were given a year to write an AI algorithm that they could then use to find and exploit vulnerabilities in a software system that they had not encountered before. It was a version of a game called 'capture the flag', in which hackers simulate real-world cyberattacks. However, the game had never before been played between fully automated systems. As Walker explains, the challenge was significant: 'You have to manage multi-party game theory,' he says, because 'it is a multi-opponent game, of incomplete information.' Unlike chess, Go and poker, it had never been won by an AI algorithm. For Walker, the challenge was a resounding success. 'When cyber-defence AI can actually work at those complexity levels, then you can see a different paradigm emerging. Autonomous cyber-defence AI is now something for which there are multiple real-world prototypes,' he tells me. We seem to be entering the early stages of a cyber arms race and from now on, the capabilities of AI systems will escalate rapidly. 'It is a bit like a human organism fighting a virus,' suggests Whitehead – 'There's always a fight going on, there's always a mutation, and unless your own immune system is adapting to that, then it will get through at some stage.' Autonomous attackers and defenders will attempt increasingly subtle and devious strategies, executing repeated rounds of attack and counter-attack within milliseconds. Walker thinks the combatants who can attract the best artificial intelligence engineers and the most powerful computers should eventually prevail, though because all sides will try to do exactly this, the end of cyber conflict is not in sight.

WAR OF THE ROBOTS

Hugh Durrant-Whyte thinks that artificial intelligence will soon start to play an even greater role in traditional warfare than it does today: 'Where you're going to see automation going in the future, is for de-manning warfare,' he says. One of the clearest examples of this is the extensive use of aerial drones for surveillance and attack; they can greatly improve the precision of air strikes, reducing the rate of unnecessary casualties.[25] The history of armed drones goes back to the First World War and 1917, when American engineers Elmer Sperry and Peter Hewitt modified a biplane by installing gyroscopes that allowed it to maintain a stable flight, follow a pre-planned route and then drop a payload of bombs onto a target. Early prototypes were plagued by technical problems, but the development of the sophisticated auto-pilot technology that now controls most commercial airplanes improved everything. The use of armed drones and robots is an increasingly common feature of modern conflicts. Drones may be controlled remotely, but although removing pilots from flying takes them out of physical danger, it does not shield them from the psychological challenges of combat. Contrary to some media reports, the evidence suggests that drones do not create a 'PlayStation mentality' among their operators; the video pictures of battle that they witness are just as disturbing as the images seen from a cockpit.[26]

Pilotless drones can now plot their own routes to targets while deftly negotiating obstacles, but they are not yet trusted to make the autonomous decision to fire a bullet or detonate a bomb. 'I design and build robots that go to war,' says Whitehead, 'but I have a rule … that we are not going to design anything that puts the decision-making about engaging the enemy in the hands of the robot. We have to have a human being in the decision-making loop.' That could change in the future, but for the moment the human decision-maker is firmly embedded in the system, and this is the position backed by the majority of experts in robotic engineering; nearly 4,000 of them have signed an open letter that calls for a ban on the development of autonomous weapon systems.[27] 'If any major military power pushes ahead with AI weapon development, a global arms race is virtually inevitable,' the letter states, 'and the endpoint of this technological trajectory is obvious: autonomous weapons will become the Kalashnikovs of tomorrow.'[28]

Hugh Durrant-Whyte and AI expert Stuart Russell fear that we could reach that point uncomfortably quickly. Russell helped to produce a highly dramatic video called 'Slaughterbots' which portrays a dystopian future in which thousands of autonomous microdrones equipped with facial recognition software and carrying small explosive charges terrorise or kill whole populations.[29] 'If I wanted to build a quadcopter that could go into a building, look for a particular person and deliver them a "package", I would be able to do that with a small team of graduate students as a course project,' explains Russell.

'The problem is much easier than building a self-driving car, because a self-driving car has to be ninety-nine point umpteen nines per cent reliable. A killing machine could be 50 per cent reliable and it'll still be useful.' This is a deeply unsettling prospect. Terrorists, Durrant-Whyte adds, are 'already some of the biggest users of drones'. Further development of armed, autonomous drones could make them as cheap, easy to use, and easy to obtain as today's remote controlled varieties. If that happened, assassins could kill with unprecedented stealth and accuracy, and unprincipled regimes could monitor and control entire communities.

Missy Cummings of Duke University acknowledges that this as possible, but is quick to point out that effective, fully autonomous killing machines still exist only in science fiction. 'The technology is not there yet,' she says. 'We still do not have the ability, whether it's in the air or the ground, to clearly recognise and make a very sure assessment that the person that they are about to kill is a good guy, a bad guy or whatever.' Even if the development of such accurate and reliable weapons is deemed desirable, the persistent engineering challenges that must be overcome to make them a reality are significant. However, it seems likely that people will try – the Kalashnikov Corporation is already reported to have developed a range of fully autonomous, ground-based robots that are equipped with powerful

FIGURE 8.8 The unimaginable terror imagined in *Slaughterbots* (2017): mass killing with face-recognition-capable drones.

FIGURE 8.9 No boys' toy: a fully-automated killer robot made by the Kalashnikov Corporation.

guns.[30] 'The horse is, kind of, bolting out of the barn door as we speak,' says Anders Sandberg, who thinks that the strong incentive to ensure that weapons of this sort work as intended offers some hope, because 'you don't want them to turn on you'. He believes that the failure of a country's military to prove that they can maintain control of these systems 'should be a war crime'. And he is completely right; developments in this area should be illegal and utterly taboo.

MILITARY RESEARCH NO LONGER LEADS THE WAY

Cummings maintains that the pace of automation development in the military sector has been slow in comparison with the great strides being made in the engineering of autonomous vehicles in the commercial sector.[31] However, this is only one of the many areas of research in which military presence is declining. The numbers bear this out: until the mid 1980s, the US Department of Defense research and development (R&D) budget accounted for 50 per cent of all such spending worldwide, but now accounts for less than a 10 per cent.[32] Commercial R&D spending within the automotive and information technology sectors have seen particularly large increases in the last two decades. Several corporations, including Google, Huawei and Microsoft, each invest well over $10 billion per year in this area. Cummings thinks that, with such huge budgets being ploughed into non-military technology, fewer of the world's most gifted engineers will opt for a career in defence R&D. Perhaps for the first time in history, the defence industry is not the primary force pushing the leading edge of technology forward; 'dual use' could return to the earlier, 'Venetian' meaning whereby flag (defence) followed trade (commerce), rather than the other way around. This may slow the development of military hardware, but it might also reduce the likelihood of the invention of technologies that could have the same transformative impact on civilian life as the Internet, GPS or the jet engine. Projects like those usually require the sort of long-term, secure funding that can only be provided by governments. Whitehead has a different concern about a future where our defences rely on more commercial innovations: 'Many of the failures in defence today are associated with taking commercial kit off the shelf and using it in a military environment,' he says. Durrant-Whyte outlines another growing risk, when he states that '99 per cent of machine learning and automation research is not happening in defence, so everybody has the ability to grab it, not just the big countries'. As ever, the technology itself is not the problem – it is rather the way that people choose to use it.

Engineering advances have always had both intended and unintended consequences, which have, in turn, produced both beneficial and harmful effects. Nowhere is this more obvious than in defence engineering, an area where deciding what we mean by 'good' and 'evil' is rarely straightforward. The world is on a trajectory of declining violence, and military deterrence is important for sustaining this trend. Moreover, most new developments in defence have civilian as well as military uses. In fact, almost all our most powerful technologies have emerged from defence-related research and development. For these reasons, governments must continue to lead the way in funding that sector. Lethal autonomous weapons are on the cusp of viability and present a new danger – at present, there are no limitations to their development, and this must be rectified. Governments must prohibit them and civil action must make them taboo, in order to prevent irresponsible players, whether states or rogue actors, from accessing them. Cyberwarfare already threatens to undermine our security and disrupt the systems upon which our lives depend – at any moment in time an unknown number of potential attackers are probing our systems for vulnerabilities. There can be no easy solutions, and the years ahead will see an escalation of increasingly sophisticated attacks and counter-attacks. Occasional security breaches and infrastructure failures are inevitable, but the future stability of our society depends upon our governments taking active measures to minimise the risk. We need heavy investment in cyber defences and back-ups in place before vital systems are compromised. And we must refrain from connecting some critical infrastructure systems to the Internet, because even as this improves efficiency, it can also make them vulnerable to attack.

While the risk of nuclear war has not gone away, the escalation of cyber and information warfare means that yesterday's doctrine of mutually assured destruction may be giving way to a threat of mutually assured disruption.[33] This is still an uneasy foundation for peace, but I believe most people would agree that disruption is far preferable to all-out destruction. In order to maintain our progress towards a less violent world, the defenders of civilised values must understand and absorb the most powerful technologies available and determine how best to use them as deterrents. As Durrant-Whyte puts it, the essential question is always the same: 'What do you really need to do to stop people being aggressive? What would prevent them, and what would make it costly for them to attack?'

Trying to understand why, in a world where so much violence and devastation is possible, so little actually occurs,[34] Anders Sandberg invokes the concept of 'Swiss cheese security'. Each new layer of defence will inevitably contain some holes, 'but if you put another slice of Swiss cheese behind it, the holes might not align. Something that could easily penetrate the first layer of defence might be caught by the second or the third, and so on.' These layers of protection are not just technological; they are also genetic and cultural. We are, for example, born with the innate capacity for compassion and empathy. As we grow older, our upbringing and education reinforces judgements about what is right and what is wrong, and a desire to maintain a good reputation and avoid retaliation helps to keep us on track. National and international regulations, if well-designed and enforced, can limit the sale of arms. The free press, police forces, online surveillance and security cameras remind us that violent acts are likely to be observed and punished. International agreements, both formal and informal, create powerful taboos that make us regard the use of nuclear, chemical and biological weapons of mass destruction as crimes that should be universally condemned. This is partly achieved by ensuring that there is always the possibility of a second strike. None of these defences will ever be impervious, but there are many layers that combine to keep most of us safe, most of the time. This layering of new defences and patching of holes in those that already exist is key to the progress of civilisation. It is all we can do to keep the would-be barbarians outside the city gates.

9

Survive

In AD 765, the Abbasid Caliphate was ruled by the second Caliph al-Mansur, the founder of a new city that we now call Baghdad, which had quickly grown to become the centre of the Arab world, a teeming trade hub and a hothouse for the world's finest intellects. His empire spread across the Middle East, occupying much of North Africa and reaching into Southern Asia. When the Caliph became gravely ill, he sent for the best physician – a Nestorian monk called Bukht-Yishu left his monastery in the mountains and crossed 150 miles of desert to reach Baghdad. Assessing his patient, he looked up at the heavens with a mysterious device called an astrolabe, which translated a detailed three-dimensional map of the night sky onto a flat surface. It was the computer of the day, and the height of sophisticated contemporary engineering. Adjusting his astrolabe according to the position of key stars in the night sky, the physician-monk could tell the exact time and calculate the precise times of sunrise and sunset, as well as the movement of many stars and planets – and because he was an astrologer, he also believed it could diagnose the Caliph's illness.

In a cluttered basement of the British Museum, the curator of horological collections lets me handle an astrolabe like the one used by Bukht-Yishu, and I am struck by its beauty. The size of a small dinner plate, it is adorned with flowing Arabic script and decorative motifs. Three interlocking brass discs and a revolving straight rule, which has a sight for measuring the elevation of objects in the sky, rotate in relation to one another with silky smoothness around a central brass pin.

To an uninitiated observer it must have seemed magical that such an apparently simple device could calculate so much with such accuracy, but that is exactly how engineered technology appears in any age – after all, very few people today understand how the Internet or the smartphone in your pocket works. What Bukht-Yishu did with his precise mathematical predictions about the stars is less impressive to the modern mind. His medical prognosis rested firmly on astrology – he believed that what happens in the skies determines what will happen on Earth. His astrolabe told him a great deal about the state of the heavens at the exact time the Caliph fell ill, and from

FIGURE 9.1 Demonstrating AI (astrological intelligence) in the thirteenth century: a Syrian astrolabe.

that he believed he could infer the root of his illness and the best course of treatment. Fortuitously, the Caliph recovered; thankfully for us all, when it comes to fighting illness we have progressed far beyond such superstitions.

In 2017, we can watch as Justin Vale, Professor of Urology at Imperial College London, slides his scalpel into his patient's exposed white abdomen that is laid out in front of him. 'What the robot can't do is make the keyhole. That is what I'm going to do now.' He quickly makes a series of small holes in his patient and then calls for the robot. An assistant wheels a bulky, six-foot-tall platform to the operating table, and four spider-like mechanical limbs are soon poised above the patient. One, holding a camera and a powerful fibre-optic light source, reaches into the largest hole, and a high-resolution picture of the inside of the patient's abdomen appears on the screens that surround the operating theatre.

At St Mary's Hospital in London, Vale is using this robot to remove a cancerous prostate gland. He sits at a console in one corner of the operating theatre from where, through his binocular eyepieces, he has a startlingly crisp three-dimensional view of what he is doing. He controls the robotic limbs, which burrow through his patient's insides. 'You can see how easy it is,' he says. Two control arms below his console are mounted on a series of swivelling linkages – he grasps them with a pincer grip and proceeds to operate on thin air 'as if I'm doing open surgery'. Each of his actions is translated into an exact, tremor-free movement of the robot's camera and the microsurgical tools.[1] We watch as the robot makes deft snips and nudges. 'There's the prostate,' Vale announces as the screens display a blue-grey mass amongst the glistening membranes. After seventy-five minutes he has cut the entire prostate free; about the size of a small lemon, the robot pulls it out through one of the holes in the patient's abdomen and places it in a specimen bag for further analysis. 'Now the fun bit: putting it all back together.' Vale uses the robotic tools to re-plumb the bladder and urethra. A series of carefully placed sutures knits the tissues back together, and the robot retreats.

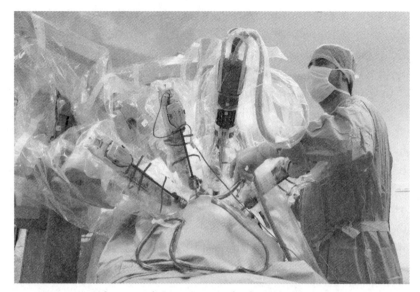

FIGURE 9.2 Robocut: a da Vinci surgical robot in action.

Vale's robotic system operates with impressive precision, but it cannot think for itself. 'Of course, if you're a purist, this isn't a real robot – it's a master-slave manipulator,' he agrees. Will robots soon be able to operate on patients autonomously, with less input from the surgeon? 'I think that is definitely feasible,' Vale replies without hesitation, adding 'whether it's desirable is another matter.'

THE POWER TO CHEAT DEATH

This surgical procedure shows the extraordinary power of medical devices today. Physicians in an intensive care unit can transfer the function of major organ systems, including the heart, lungs, kidneys and liver, to a mechanical device. Clinicians can distinguish between healthy and damaged tissue using magnetic resonance imaging (MRI). Surgeons can conduct longer, more invasive and life-saving procedures, including organ transplants, secure in the knowledge that antimicrobial drugs will keep infection in check. As a result of these innovations, multi-organ failure, cardiac arrest and septic shock no longer mean an immediate death. Some ambitious scientists now want to stave off death entirely, by fending off infectious diseases, updating genetic operating systems and making replacements for worn out cells and organs. This remains a fanciful ambition, but the fact that any scientifically trained person is able to talk in these terms shows just how far and how quickly medicine has progressed.

It is only relatively recently that a child could be born with a realistic prospect of surviving into old age, let alone avoiding having to live out their final decades in infirmity and frailness. People now live for an average of seventy-two years; at the beginning of the twentieth century, average life expectancy was just thirty-one.[2]

For most of our history we were powerless in the face of all but the most trivial illnesses. Chronic diarrhoea and wounds that became septic cut short the lives of countless millions. Childbirth was a dangerous ordeal for mother and baby alike. Terrifying plagues swept across continents, killing as they went. Physicians did not usually

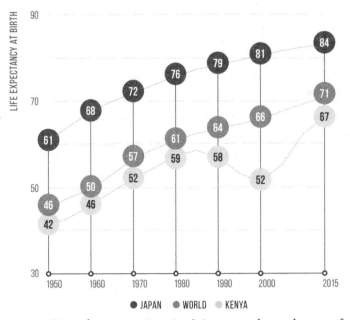

FIGURE 9.3 Living longer, getting wiser? Are we ready to take care of new challenges in health care?

understand what was causing an illness, and their attempted cures often did more harm than good. Despite the lack of any evidence of their effectiveness, bloodletting, trepanning and the application of burning-hot cups to the skin were used for thousands of years, in both Eastern and Western medicine. During the nineteenth century, health care slowly started to change from a hopeful enterprise based on superstition and speculation, into a rational programme of diagnostic, preventative and curative measures to extend and improve lives. Scientists and engineers built tools that allowed them to learn about the real causes of disease and the workings of our bodily systems. These great achievements, together with advances in public health, infrastructure, housing, education and diet, relieved humanity from millennia of endemic disease and famine.

However, improvements in medical care and public health have created a new set of challenges, because today most lives are ended by conditions associated with old age – of the top ten causes of death in wealthy countries in 2016, just one, pneumonia, was directly linked to an infectious disease.[3] All the others, including cardiovascular diseases, various cancers and Alzheimer's disease, are chronic conditions that are prevalent in the elderly and have complex causes that involve some combination of genetic predisposition, lifestyle and environmental factors, such as obesity, smoking or exposure to pollution, as well as the accumulated wear and tear of old age. As people live longer and their diseases are diagnosed more accurately, the costs of health care will rise. While it is true that, according to UK brain surgeon Henry Marsh, 'you could spend the entire national income on health care but everybody would still die,'[4] it does not mean that we can slow progress. Rather we must become even more efficient.

The biomedical scientist and entrepreneur Craig Venter questions the whole basis of contemporary health care, which he describes as 'medieval'. 'If people have no obvious symptoms and they look OK, they are deemed by the medical system to be healthy,' he says. Venter, together with many other researchers and physicians, strongly believes that we should be more able to spot the earliest signs of disease and predict future problems so that we can intervene before they develop – today's medical practice is typically *re*active, when it should be much more *pro*active. I find this a compelling idea; watching Professor Vale and his robot perform surgery was an engrossing spectacle, but it would be so much better if such dramatic interventions were only needed *in extremis*. People would be healthier if we could do more to keep them out of hospital, and costs would be reduced. This is not a new idea – the first extensions of life resulted from preventing disease, rather than from treating it.

THE SANITATION REVOLUTION

Four epidemics of cholera swept through industrial Britain during the nineteenth century. If you were unlucky enough to catch the disease, your chances of dying were close to 50 per cent. This singularly unpleasant condition 'liquefied a body as fluids streamed uncontrollably and insensibly from both ends'.[5] While most infectious diseases affect the old, the young and the chronically sick, cholera can affect anybody. It was the need to stop this terrifying disease that inspired the public health movement, which was grounded in the simple idea that illness was a *cause* rather than a result of poverty. Cholera, with its ability to incapacitate or kill even the most able-bodied of workers, provided the clearest evidence that, if the UK government could keep the nation healthier, then workers could work their way out of poverty and boost the country's economy. As to the source of diseases, most believed they were caused by the 'miasma', invisible poisons carried in polluted air. This theory seemed to fit with what people saw around them; the excrement of London's 2 million inhabitants was hard to avoid, and many were packed like sardines into slum dwellings. Their cesspools overflowed, venting noxious liquids and smells, and graveyards overflowed with barely buried and rotting corpses. Manure and household waste clogged even the most respectable of streets, and the pollution of industry enveloped the city.[6]

The first organised plans to improve public health did more harm than good, however[7] – London's first primitive sewerage system used the city's original networks of small rivers and creeks, which

FATHER THAMES INTRODUCING HIS OFFSPRING TO THE FAIR CITY OF LONDON.
(*A Design for a Fresco in the New Houses of Parliament.*)

FIGURE 9.4 A foul river, fallen foul of sanitary standards: the River Thames (*Punch* magazine, 1858).

all flowed into the River Thames, leaving it poisoned and foul. Disastrously, several water companies drew water from the river and sold it to the population. When cholera returned once again in 1854, the pioneering epidemiologist John Snow, now well known as the person who showed that cholera is carried in water rather than in the air, built up a body of evidence to disprove the existence of miasma. However, he was pushing against received wisdom about the causes of disease, and it was only later that his huge contribution was fully accepted and celebrated.[8] June 1858 was unusually hot, and the polluted Thames gave off 'the Great Stink of London', which caused the evacuation of Parliament. The first wave of sanitary reform had failed.

Something bigger and bolder had to be done. The increasingly desperate government latched onto the ambitious plans of Joseph

Bazalgette, an engineer who, during the subsequent decade, oversaw the construction of a robust arterial system of sewers consisting of over a thousand miles of brick tunnels. These sewers absorbed London's sewage and rainwater and conveyed it away from the city.[9]

The scheme was clearly a success – cholera never returned,[10] and the incidence of typhoid, dysentery and water-borne infections plummeted immediately. Bazalgette's obituary in 1891 concluded that 'the great sewer that runs beneath Londoners ... has added some twenty years to the chance of life'.[11] Although they are 150 years old, his precisely laid sewers are still in excellent condition and the health of London still depends on them, even as they struggle to cope with the city's expanded population;[12] his impressive Victorian engineering was clear evidence of the effectiveness of organised public health initiatives.

London's sewers inspired many other similar engineering projects around the world. In Chicago, for example, an immense civil engineering project in 1900 reversed the flow of the Chicago River, which successfully dealt with the pollution and sewage created by the city's

FIGURE 9.5 No time to waste: Bazalgette's construction of London's great sewers dealt with the Great Stink in the 1860s.

booming industrial economy. However, the sanitary revolution that Bazalgette started is still incomplete; in 2009, around 60 per cent of the world's population still lived without access to effective wastewater treatment facilities.[13] Cholera still kills tens of thousands of people every year, a collective failure that could readily be fixed; although large-scale municipal facilities are expensive to build and maintain, the value of productive lives saved should be factored into the costs. In any case, cheaper solutions, including pit latrines that can be safely emptied, already exist and could be applied much more widely, along with better public health education.[14] More encouragingly, however, over 90 per cent of people now have access to improved water sources, and that figure is rising quickly.[15]

FIGHTING GERMS AND FINDING VACCINES

John Snow was not the only person working to understand the causes of infectious diseases in the nineteenth century. A small group of investigators, notable among them Louis Pasteur in France and Joseph Lister in the UK, were developing 'germ theory', which proposed that infections are caused by invisible life forms such as bacteria, viruses, single-celled fungi and protozoa. Crucial to their progress was the ability to see these 'germs' and describe them to a sceptical public. In vastly improving the resolution and image quality of the compound microscope, Joseph Lister's father, an optical instrument engineer, provided them with a crucial tool. Lister was able to use his growing knowledge of microorganisms to develop and promote safe, antiseptic practices. He discovered the power of carbolic acid, an extraction from coal tar, in the summer of 1865, and began wrapping the surgical wounds of his patients with bandages soaked in this strong antiseptic. The results were impressive – the proportion of people dying after his surgical procedures dropped from 50 per cent to 15 per cent almost immediately.[16] The use of antiseptic and aseptic techniques, including the critical realisation that physicians must wash their hands,[17] soon spread to other areas of medical care. When they were used during childbirth, infant and maternal mortality rates fell dramatically.[18]

Much of Lister's inspiration came from Pasteur, the French scientist whose investigations into microscopic organisms led him to invent the heat treatment technique now known as pasteurisation, which improved the stability and safety of dairy products and many other foods. Even more important was Pasteur's realisation that killing or reducing the virulence of bacteria and viruses could turn them into effective vaccines – he created vaccines against anthrax, rabies and chicken cholera, and his techniques were used by others to create or improve several other vaccines. Completing the earlier work of Edward Jenner, smallpox was finally eliminated in 1980, and polio may be next. Vaccines also ensure that several other important diseases, including diphtheria, whooping cough and measles, are well contained in most parts of the world. A number of engineering advances mean that pathogens can be characterised and a vaccine can be designed, mass-produced and shipped with impressive speed.

For example, at the start of the Ebola outbreak in West Africa in 2014, there was no effective vaccine available – within weeks, several

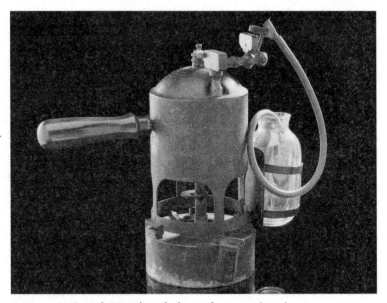

FIGURE 9.6 Joseph Lister's carbolic acid sprayer (1867).

new experimental vaccines were being tested for safety and efficacy. By the end of 2016, the first of these was shown to be safe and effective at containing the spread of the disease;[19] it had arrived too late for the 11,000 people who had died and the further 17,000 people estimated to have been infected,[20] but there is now at least one effective vaccine to apply as a defence mechanism in future outbreaks.[21] Although advances of this kind are significant, doubts about the safety and effectiveness of many vaccines are widespread. The demonstrably false claims made by the English doctor Andrew Wakefield about the links between the combined measles, mumps and rubella (MMR) vaccine and autism may be the clearest example of this belief, but irrational scepticism about vaccines is now common, in rich and poor countries alike. Concerns must be answered sympathetically, but we must not allow them to threaten the great strides we are making towards the elimination of many infectious diseases.

Jeremy Farrar, director of the Wellcome Trust, applauds this dramatic scientific progress, but also points out that our world is still too vulnerable to disease pandemics. He reminds me of the influenza outbreak of 1918, which infected nearly a third of the world's population and killed between 30 and 50 million people. 'Influenza is arguably the one infection we know about which could bring the world to a complete standstill – it is unacceptable that we have a seasonal flu vaccine that is only 30 to 40 per cent effective, takes six to twelve months to make, and has to be made in eggs essentially using 1950s technology!' Thankfully, things are changing. 'I think a universal flu vaccine is scientifically feasible,' says Farrar, rejecting the idea that flu strains are too varied and change too quickly for a single vaccine to provide protection. Indeed, several laboratories around the world are now taking what Farrar calls 'an engineered approach' to developing an effective flu vaccine.

ANTIBIOTICS: A FRAGILE GIFT IN THE FIGHT AGAINST DISEASE

The prevention of bacterial infections through sanitation, antiseptic procedures and vaccination has been transformative, but just as important is our ability to destroy these infections if they manage to outwit these defences. Throughout most of history this was impossible: there were no effective drugs for treating bacterial infections. The first was penicillin, which was serendipitously discovered, in 1928, by Alexander Fleming when he found a secretion from a blue-green mould that could kill bacteria.[22] However, after its discovery, only one person was treated with it during the subsequent fourteen years, since it was very difficult to extract from the mould in which it was generated. Only after US engineers, driven by the increased wartime demand for antiseptic, turned their attention to the problem could it be extracted in useful quantities. One official summarised the challenge: 'The mould is as temperamental as an opera singer, the yields are low, the isolation is difficult, the extraction is murder and the purification invites disaster.'[23] However, the US Department of War had noticed penicillin's potential and made achieving its production a priority second only to the Manhattan Project.[24]

A chemical engineer called Margaret Hutchinson Rousseau tackled the challenge head-on. Having gained a reputation devising ways to produce synthetic rubber and to distil oil into high-octane airplane fuel, she applied that thinking to the penicillin problem. Her innovation was to grow the penicillin-producing mould in large quantities, in deep tanks.[25] Engineers worked flat-out trying to get the penicillin mould to grow, setting up vast steel fermentation tanks and feeding them with slurry made from corn steep liquor, which they enriched with extra sugars and minerals. A complex set-up of pipes and stirrers bubbled and agitated the mixture, letting the mould breathe. After several days of brewing came the delicate task of extracting the active ingredient and freeze-drying it for storage. By the time of the Normandy beach landings in June 1944, the Allied forces had 2.3 million doses of penicillin, most of which had come from fourteen giant fermenters in a converted Brooklyn ice factory, each of which held 28,000 litres of penicillin mould. A year later, they ramped up production to a huge 650 billion units per month – 1945 was also the year in which Fleming, along with Howard Florey and Ernst Chain, the Oxford scientists who confirmed and expanded his basic findings, received their Nobel Prize. While the pomp and ceremony unfolded in Stockholm, Margaret Hutchinson Rousseau, a female engineer rather than a male scientist, was at home looking after her young son.[26]

Several new classes of antibiotic were isolated in the following years; as a result, the foul smell of gangrene no longer lingers in hospital wards, and the secondary infections that so often turned minor operations into life-threatening ordeals are rare. Even the bubonic plague, which killed nearly half of the Chinese population and a third of all Europeans in the middle of the fourteenth century, is now treatable with antibiotics.

Antibiotics may have helped save many lives, but their indiscriminate use has created a grave problem. In 2016, a seventy-year-old woman died in the intensive care unit of a hospital in Reno, Nevada. Her doctors knew exactly what had killed her – it was an infection caused by a relatively harmless species of bacteria called *Klebsiella pneumonia* – but they could do nothing about it.[27] Her physicians had tried the fourteen antibiotic drugs at their disposal, but all had failed. At the federal Center for Disease Control in Atlanta, Georgia, scientists also tried every other antibiotic licensed for use in the US, but none of them worked. The 'superbug' rampaged through the woman's body and sent her into septic shock. Her death was the first US case of a lethal bacterial infection that resisted the entire arsenal of approved antibiotic drugs. The appearance of such drug-resistant bacteria had long been expected – antibiotics have for decades been prescribed and used indiscriminately.

'My view is that antimicrobial resistance is as complex a problem as climate change, or more complex,' says Sally Davies, Chief Medical Advisor to the UK government. 'But actually, we'll all be dead if we don't address it, because we'll have lost modern medicine well before climate change does us in.' Davies points out that the challenges posed by climate change and antimicrobial resistance have similar features; both started as invisible problems that have since became global in their scope, with the potential to cause devastation and great suffering. This is not a new issue and Fleming warned about it during his Nobel acceptance speech in 1945, saying, 'There is the danger that the ignorant man may easily underdose himself and, by exposing his microbes to non-lethal quantities of the drug, make them resistant.'[28]

Once again, we see that however potent an engineered advance is, society must use it wisely, and to do that requires wide-ranging vision, effective education and sensible regulation. The economist Jim O'Neill has calculated that, within a generation, bacteria that are resistant to known antibiotics could kill 10 million people every year.[29] That could cost the global economy $100 trillion, equivalent to more than five times the 2016 GDP of the US. Millions would carry unshakeable infections, and hospitals would become places for the vulnerable and the sick to avoid. None of this is an appealing prospect, but Davies believes that it is not too late for such a disaster to be averted. It will need strong leadership and investment to both scale up efforts to improve diagnosis and surveillance, and to create new materials that resist infections for use in catheters, dressings and in other medical equipment. And, just as Margaret Hutchinson Rousseau did in the 1940s, engineers will do the unglamorous but crucial work of turning promising experimental cures into mass-produced drugs that are safe and stable. We have the ingenuity to solve the problem of antimicrobial resistance, but we also need the funding and political will to help us take decisive action.

IMAGING GIVES A NEW PERSPECTIVE

As infectious diseases become less frequent causes of death, a range of complex, non-communicable diseases are taking their place, many of which are exacerbated by the normal ageing process. They tend to start gradually and become progressively more debilitating; our challenge is to predict their course and take steps to avert or limit the damage as early as possible. A crucial step is the ability to see what is happening inside the body. Magnetic resonance imaging (MRI) scanners generate extraordinarily detailed images, which reveal the fine details of cartilage, bone, muscle and sinew, allowing physicians to identify the effects of injuries. Clusters of tumorous cells can often be spotted long before any obvious symptoms of cancer are visible.

Scientists studying the brain use MRI to watch the patterns of activity that ripple through it as we think and act, and are able to identify the damage associated with dementia. MRI is harmless, non-invasive and does not expose the patient to radiation – its images are generated by manipulating the quantum properties of the atomic particles that make up the human body.[30]

I meet engineer Simon Calvert at a factory that builds MRI scanners, where he explains the manufacturing challenges involved with the technology. At the core of the scanner is a very powerful magnet, the field of which must be entirely uniform and stable. The only way to build such a magnet is to use superconducting coils, each of which is made up from one hundred kilometres of niobium-titanium alloy wire, pressure-reinforced with epoxy resin and encased in a steel vessel. The coil is then cooled, with liquid helium, down to 4.2 degrees kelvin, or minus 269 degrees Celsius. An astonishing change then takes place, as the wire becomes superconductive and loses all electrical resistance. 'You can put enormous current densities into the wire,' explains Calvert, 'and then disconnect the power supply.' The huge current then circulates indefinitely, generating the very strong and steady magnetic forces that are needed for the MRI scanner.

During a scan, the patient is only a few centimetres away from the extreme cold, high current and pressure of the magnet, so needs to be protected. The magnet also needs protecting since the smallest amount of heat, even tens of microjoules,[31] would destroy its super-conductive state. The most powerful MRI machines approved for use in hospitals now work at seven teslas, which is about 140,000 times stronger than the Earth's magnetic field, and strong enough to turn unsecured magnetic objects into dangerous projectiles.[32] This strength improves the resolution of MRI from three millimetres to about half a millimetre, around the width of a grain of sand, allowing physicians to detect ever more subtle anomalies within our bodies.

The promise of being able to diagnose problems early and of monitoring the effectiveness of treatments is driving up demand. In 2013 Craig Venter founded Human Longevity, Inc., the mission of which is 'revolutionising human health by generating more data and deeper understanding into what can keep you living healthier longer'. The company offers a range of genetic, biochemical and imaging tests to assess the physical health of their customers, including a detailed whole-body MRI scan. Venter sees this procedure as transformative, because it can sometimes diagnose diseases sufficiently early that they can be easily cured. He describes how 'many people, thinking they are free from disease, come in ... and about 40 per cent of them are diagnosed with at least one very serious condition'. Venter had the scan himself and was diagnosed as having high-grade prostate cancer. He had surgery almost immediately.

Some physicians have concerns about doing MRI scans on apparently healthy people. Oxford University's John Bell believes that scans are 'really, really noisy ... and about 20 per cent of normal people have little white spots in one bit of their body or another, which are probably inconsequential but come up on an MRI scan'. Darzi has similar concerns about the risk of false-positive diagnoses: 'There is very strong evidence that people have ended up in wooden boxes because interventions were done unnecessarily.' Venter counters this by arguing that misinterpretation of MRI scans is 'really an old issue', and claims that modern techniques and growing datasets mean that 'tumours light up like a light bulb'. I agree with Venter; it is better to know that something is wrong rather than live in blissful ignorance.

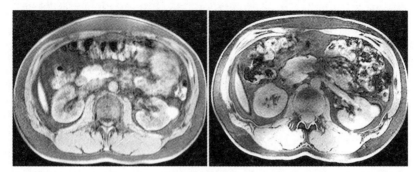

FIGURE 9.7 MRI – the detail is in the tesla: a 1.5 vs a 7-tesla system (*left* to *right*).

That, however, does mean that diagnostics need to be sufficiently accurate.

York University's Professor Simon Duckett has a rather different concern about the way MRI is being developed: he worries that the current trend towards bigger scanners and ever more powerful magnets is inexorably driving up costs, thereby limiting access to this powerful technology. Duckett is instead leading advances in a technique called hyperpolarisation, which can render visible molecules that are usually invisible. While more traditional applications of MRI rely on the inherent magnetic properties of water molecules, the method he is developing magnetises a wide range of biological molecules before they are introduced into a patient's body and imaged. 'In theory there is the potential to improve sensitivity by 200,000 times at the 1.5 teslas field of a routine hospital scanner,' Duckett says. With such a dramatic increase in sensitivity, the power of the magnets needed for imaging could be greatly reduced. 'I could show you measurements where we can detect a response simply using the Earth's magnetic field,' he explains. He foresees a time when surgeons could use this technique during an operation to monitor patients in real time, rather than having to put them in the tube of a huge scanner. The magnetic field could be instead provided by a 'wand' that doctors could pass over the area being operated on; such technology could, for example, provide them with a way to visualise and remove clusters of cancerous cells that are otherwise indistinguishable from their healthy neighbours. 'Think about operating on a brain tumour,' Duckett says. 'The major challenge there is, if you take out too much tissue, you impair function, but if you leave tissue that's cancerous, you impair life.' Duckett outlines many other potential applications of his hyperpolarisation technique, including ways to monitor the chemical composition of urine and other bodily fluids, in order to track symptoms and monitor drug efficacy. 'I think if we're talking about this as a platform technology ... it's quite stunning,' says Duckett. However, for the time being, his ideas have not yet found their way into commercial products. 'We've basically been told it's too disruptive,' he says. 'You go and talk to one of the big players in MRI technology, and you're suddenly challenging their cost model.'

There is a strong drive to make imaging more sensitive, while also reducing its cost so that it can be used more frequently. However, to make the results more informative and less prone to misinterpretation, they must be combined with data from many other sources. In particular, important clues to the future course of disease can come from understanding what is written in our genes.

CRACKING LIFE'S CODE

In June 2000, US President Bill Clinton and the British Prime Minister Tony Blair announced the first complete survey of the twenty-three pairs of chromosomes that make up the human genome. Clinton praised it as 'the scientific breakthrough of the century, perhaps of all time … it will revolutionise the diagnosis, prevention and treatment of most, if not all, human diseases'. I vividly remember the excitement that surrounded that event – the hope was that we would learn to predict our future health with great accuracy and take rational action when necessary. But, nearly two decades on, I wonder what has happened to the promised revolution in health care. My physician does not yet base his prescriptions on my genetic profile and has expressed no interest in sequencing my DNA, though it is clear that the availability of data on DNA sequences is no longer a limiting factor. 'There is probably no sector of engineering that has changed by more orders of magnitude than DNA sequencing,' says Craig Venter. The speed of sequencing has greatly increased, even as the cost has plummeted.[33] This is probably the only technology that has advanced more quickly than Moore's Law. Sequencing the first genome took over a decade and cost nearly $3 billion. Today, by contrast, you can spit in a tube, send it off, and get your full genome sequence back within a couple of weeks, at a cost of less than $1,000.

Clive Brown has spearheaded the development of one of the most exciting new approaches to DNA sequencing, the MinION sequencer. At its core is a thin synthetic membrane shot through with thousands of tiny pores, and immersed in a liquid-filled chamber. A voltage difference across the device impels the long, string-like molecules of DNA being sequenced to travel through the 'nanopores', and as they thread their way through, they alter the flow of current. Each of the four letters of the genetic code, A, T, C and G, causes a subtly different deflection in the current and by learning to recognise these signals, the MinION can read very long strands of DNA in a single process, in real time.[34] To make their new approach work, Brown had to explore using new materials, tinker with numerous proteins from bacterial cells and push the limits of microelectronics fabrication. While he was developing this new technique, many other scientists doubted its feasibility: 'Nobody thought it was possible,' he says now. Most DNA sequencers used today are fridge-sized contraptions, so I am surprised when Brown produces a sleek metal device, just four inches long and one inch across. It is straightforward to use, plugging into any computer with a USB cable. With tools like this, obtaining DNA sequencing data is easy. Understanding what the data means, however, is another problem altogether.

Professor Paul Freemont, a synthetic biologist at Imperial College London, uses an analogy to explain the problem of interpreting genome sequence data: 'It's a bit like taking a Boeing jet apart and having all the little components laid out in a big warehouse.' Sequencing is a crucial first step as it provides us with the template of the organism's genetic code, but we are still a long way from knowing how all the elements that are produced by the code fit together to form a living cell and a whole organism.

The human genome is 3 billion letters long. If you are trying to find out which genetic factors make you unique, more than 99 per cent of those letters are uninformative; this is our common genetic signature, identical for every person on the planet. Only a fraction of the remaining 1 per cent, which still amounts to several million bits of code distributed throughout our genomes, varies. Geneticists call these variants; some of them explain why my eye colour is different from yours, while others influence my unique set of tastes, physique, mental habits, health risks and which pharmaceutical drugs my body will respond to. Many variants will only have an effect in specific circumstances; a certain variant may increase your risk of developing cancer, for example, but only if you are a heavy smoker. Many others have no discernible effect at all.[35] If the combinations of genetic variants that are the most predictive of threats to your health could be identified, you could structure your life to avoid unnecessary risks and look out for the very earliest signs of predicted illnesses. As a result, health care could become more proactive and personalised to your specific genetic make-up.

This is an appealing strategy, but implementing it is proving a major challenge. Variants that have big effects on an individual's susceptibility to disease are rare. Mark Walport explains that 'it turns out that most diseases, like the common causes of diabetes and heart disease, for example, are under very, very polygenic control'. Small changes to large numbers of genes nudge our predisposition to these ailments by tiny degrees; genomics is a world of probabilities and conditionalities, and unpicking cause from correlation remains a stubborn problem. 'The reason we don't know how to do that is because nobody's ever done the key experiment,' explains John Bell, 'which is to look at a lot of normal people, find those variants and then follow them over long periods of time to see whether the experimental participants get any trouble.' Then Bell introduces me to someone who has launched just such a project.

Professor Rory Collins leads UK Biobank, a long-term study of half a million adults. As well as having their genomes surveyed, participants consent to a barrage of blood tests, cognitive assessments and brain and body scans, and also agree to share their health records. The aim is to put genetic information in a meaningful environmental context. The dataset that the project is generating is a treasure trove for researchers. As Collins explains, 'the problem will then be how anybody can pull out the signals.' He does not yet have the answer, but the Biobank's data is accessible to anyone for approved research and Collins is particularly keen to engage the data engineering expertise of 'the kind of the people who are creating Silicon Valley'. The benefit will be mutual though, as 'The Googles of this world haven't had access to these kinds of really high-quality, large-scale databases to identify ways to improve health.' Insights gained from the study of genomic databases by experts from many different fields will deliver many new drugs and take us towards the sort of future promised by Clinton and Blair, where our genomes are used to make health care more proactive. Despite the progress already made, economic constraints and difficulties of scientific interpretation will limit how far treatments can be personalised. What is already happening for a range of conditions, from HIV to malaria and several cancers, however, is the classification of patients into groups that will respond to particular drugs according their genetic make-up. This is improving care and cutting costs.

DATA: THE NEW MICROSCOPE

We do not yet live in a world where chronic threats to our health and well-being are routinely handled at an early stage, in spite of the great progress made with such powerful devices as the MRI scanner and the DNA sequencer. To reach that goal we need to get much better at integrating multiple sources of information and moving fluidly between diverse fields of expertise. John Savill, formerly the Chief Executive of the UK Medical Research Council (MRC), believes that big data, which he calls 'the microscope of the twenty-first century', will eventually have a huge impact on health care. His top scientific priority is 'to see the integration of all of the different types of clinical and biological data that is gathered from humans'. We will also need to gather and integrate information on dietary habits, addictions, exposure to pollutants and so on, since almost all chronic conditions arise from a complex mixture of not only genetic influences but also environmental ones.

There are encouraging signs that this convergent approach to health care innovation is starting to gather momentum. It is what Craig Venter in the US, the UK Biobank project and several other high-profile international projects in China, India, Mexico and beyond all aim to do.[36] These projects are all amassing large datasets and using them to identify the strongest predictors of good and bad health. To succeed, they will need access to high-quality and well-organised data. There is a quite significant risk that widespread concern about the infringement of personal privacy will limit data sharing, as I discussed in Chapter 4. However, as Bell makes clear, 'we'll never find out about the most important causal genetic variants, unless everybody shares their genomes'. Repositories of medical data should enforce rigorous security and anonymisation; and wherever possible they must be accessible to all researchers so that people do not feel their data is being used by others for inappropriate financial gain. The increasingly proactive health care systems of the near future will need a new generation of therapeutic tools to conduct ever more precise interventions, right down to changing single letters of DNA code.

Technologies will converge. In surgery, for example, the current generation of robotic assistant has done more than just improve procedures. Ara Darzi thinks that the benefits are much greater, since 'what robotics truly did is brought the microchip into the operating theatre'. This innovation allowed surgeons to use information from MRI and computer-aided tomography (CAT scans) to guide their operations. He explains how robots can move to accommodate the breathing of the patient, creating a stable platform, asking, 'How else are you going to stitch a tiny six-millimetre blood vessel while the heart is beating? Suddenly you can, and that is engineering.' He describes yet another advance, the iKnife,[37] which analyses the chemical composition of the cells that it vaporises as it cuts. It can learn the difference between normal and cancerous cells, telling the surgeon when, for example, she has removed the farthest margins of a tumour in a breast. With these new tools, surgery will get more exact and reliable, and less traumatic for the patient's body. He also believes that surgeons may well soon be able to trace around a virtually rendered tumour, before letting the robot conduct the operation. The prospect of robots performing surgery autonomously may make some uneasy, but Darzi is convinced that medical staff will only continue to be able to provide adequate care if they actively innovate and adopt all the technological aids available. 'The ageing demography just expands,' he says, leading him to conclude that 'the burden of disease is so big that we're never going to be able to produce enough practitioners'. It seems that physicians will need all the help they can get from robotics, artificial intelligence and other maturing technologies.

MODERN DRUGS: IMPROVED DELIVERY AND TRANSFORMATIVE TREATMENTS

At MIT I meet an engineer who demonstrates like no other the power of this new, convergent approach to biomedical engineering. Every inch of available wall space in Robert Langer's large office is covered by framed certificates, commemorating his many awards. Described recently as 'the Edison of Medicine',[38] he has over one thousand patents, both granted and pending, and has published well over a thousand scientific papers. The thirty companies that he has helped to launch have a combined market value of $25 billion, and only one of these ventures has failed. Langer is probably the world's most influential engineer, but despite the abundant evidence of his eminence, he is refreshingly down-to-earth and humble about his achievements. The focus of his work is deceptively simple. '[Drug] delivery is probably the number one thing,' he says. The most potent medicines in the world can be effective only if they are able to reach the parts of the body where they are needed, in the correct state and at the right time – this is what he is trying to achieve.

Langer's large laboratory is populated by chemical, mechanical and electrical engineers, materials scientists, molecular biologists, physicians, veterinary scientists, physicists, chemists and computer scientists. Rather than staying within the confines of their own disciplines, they work productively and side by side, finding the shared ground required to make progress. Time and again, Langer and his diverse team have demonstrated their ability to imagine novel strategies for delivering drugs, turning these ideas into practical strategies that work at scale. Among the many innovations that Langer describes are nanoparticles, just a few billionths of a metre across, that specifically seek out cancer cells and deliver drugs to them. He has also developed polymers that release drugs in a slow, controlled fashion over weeks or months, and microchip-based devices that release their payload only when stimulated to do so by remote control. Recently he developed a form of modified table salt that could treat the vitamin and mineral deficiencies that afflict many of the world's poorest people.[39]

I was particularly struck by one of Langer's inventions that demonstrates how the best engineering can slice through complexity and inefficiency to deliver an imaginative solution to a serious problem. The oncologist Siddartha Mukherjee recalls his daily struggles to devise effective and humane attacks on cancer, which he calls a 'grotesque and multifaceted illness'.[40] 'Was it worthwhile continuing yet another round of chemotherapy on a sixty-six-year-old pharmacist with lung cancer who had failed all other drugs?' Mukherjee ponders. 'Was it better to try a tested and potent combination of drugs on a twenty-six-year-old woman with Hodgkin's disease and risk losing her fertility, or to choose a more experimental combination that might spare it?' As Mukherjee explains, finding the best chemotherapy agents is often a drawn-out and painful experience, because there are over a hundred drugs to choose from.[41] Usually physicians have to apply them sequentially, over periods of weeks, using indirect tests to help them decide if the drugs are working. What is more, most chemotherapy agents are poisons, and their side effects can be atrocious. 'If we didn't kill the tumour, we killed the patient,' wrote leukaemia pioneer William Moloney in 1997, describing the early days of chemotherapy development.[42]

However, a new device from Langer's lab could soon eliminate much of the guesswork and agony from this process. 'You do an experiment that tells you what chemotherapy regime you should use … You actually bring a laboratory right into the patient for a day,' Langer tells me. The 'laboratory' to which he refers seems surprisingly simple. There are no moving parts, no complex electronics, nor advanced nano-machines at work here – it is little more than a tiny rod, slightly smaller than a grain of rice, made from a robust resin. Micro-machined into it are dozens of tiny reservoirs, each preloaded with a different chemotherapy drug. The device is sent down a needle

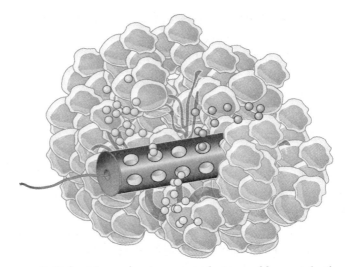

FIGURE 9.8 Robert Langer's microscopic device could soon take the guesswork out of cancer care.

into the body of a tumour, where the anti-cancer drugs then seep out into the cancerous tissue. With an engineer's confident grasp of physics and mastery of materials, Langer is able to control the size and chemical properties of the drug reservoirs. The next day, a slightly larger needle recovers the device and a halo of tumour cells surrounding it. Technicians then scrutinise the patches of drugged cancer cells, and are able to see immediately which drug cripples or kills the tumour cells best. 'We're in clinical trials right now; that moved surprisingly quickly,' Langer told me in 2017. If these tests work out, his device could radically change cancer care. The first version of the device had sixteen reservoirs, but 'now it can take up to a hundred different drugs'. As well as transforming drug selection and dosage calculation, Langer believes his tool will also speed up the development of new cancer drugs. This is because, at present, many candidate pharmaceuticals fail at the final hurdle of the clinical trial, when compounds that work in the controlled conditions of laboratory tests and animal experiments fail to work as expected in the complex ecosystem of the human body. At that stage, hundreds of millions of dollars have already been invested. Langer's device would allow low-risk testing, in a patient, at much earlier stages of drug development.

Does it take the incisive mind of a leading engineer like Langer to innovate in this way? 'A lot of scientists have done some wonderful, wonderful work when they focus in on a single bit of genetics or a microorganism their whole life,' he says, 'but I think engineers are broader. We see a problem and just say, "Well, maybe we can attack it differently."' Part of his success stems from his ability to develop 'platform technologies'. 'I think most really good engineering technologies have many applications,' he explains. The drug delivery vehicles that he has designed and built carry many different pharmaceutical cargoes into the body. In doing so, they have saved hundreds of millions of lives and improved the quality of billions more.[43]

Transformative as so many of Langer's inventions are, his ideas have not always been warmly received by the medical community. Back in the 1970s, when he was a junior professor, his first nine research grant applications were turned down. This setback would have derailed many careers, but Langer was used to fighting his corner; he had applied for forty different assistant professor jobs before finally finding, at Boston Children's Hospital, the only professor open-minded enough to welcome an engineer into a medical school. Since then, Langer has repeatedly cracked problems previously deemed unsolvable by the medical establishment.

While it is often said that engineers save more lives than doctors, Langer's work shows just how far beyond the provision of clean water and sanitation the engineer's impact can go. Every pill and every device has also been engineered. Engineers, therefore, play some hand in every single medical success story, without exception. I ask Langer which drugs he thinks hold most promise for the future. In his opinion, some of the most potent medicines will be those that change the activity or the sequence of our genes. 'I think they will be revolutionary,' he says.

VISIONS OF LIFE WITHOUT DEATH

Biologists have been developing ways to modify and edit DNA since the 1970s. Until recently, this work was frustratingly laborious and hard to control. However, CRISPR-Cas9, an enzyme system found in many species of bacteria, is changing that, by removing much of the 'black magic' from DNA editing. As a result, it is helping biomedical scientists to think and act like engineers. It has only been available for use on mammalian cells since 2012, but already provides a vastly accelerated and reliable way to change the sequence of DNA, thus correcting errors and adding new genetic functions. The first wave of therapies that rely on CRISPR-Cas9 will soon enter the clinic.[44] In particular, a new generation of cancer immunotherapies, treatments that adapt or augment a patient's own immune system to seek out and attack cancerous cells, have great promise. In clinical trials, chimeric antigen receptor T-cell (CAR-T) therapy has demonstrated an ability to effect dramatic recovery from advanced blood cancers. This approach involves extracting immune cells, or T-cells, from a patient's blood, modifying their genomes with CRISPR-Cas9 and re-engineered viruses to make them find and stick to cancer cells, multiplying them in the incubator, and then injecting them back into the patient, where they have a potent ability to eliminate their cancerous targets. Since these genetically modified T-cells are recognised by the body as 'self' they are not rejected by the immune system. CRISPR-Cas9 is now also used to make further genetic changes that reduce side effects and ensure that the modified T-cells cannot destroy non-cancerous cells.[45] CAR-T therapy is the first of a new set of rationally conceived, highly precise cancer therapies and is currently receiving regulatory approval in several countries. John Savill hopes that similar approaches will then be applied to a wider range of diseases and conditions – he is optimistic that 'within a very few years, there will be genetically engineered stem cells being put back into patients with a disorder in a particular organ'. These stem cell-based treatments would not create inheritable changes. 'It's for *your* body – it's not what your children will inherit,' he explains.

However, some researchers are keen to make more permanent changes – so-called 'germline gene editing'. For instance, CRISPR-Cas9 could correct the aberrant DNA variants that cause a range of inherited diseases, including cystic fibrosis and muscular dystrophy. In principle, the same technique could be used to create resistance to viral infections such as HIV/AIDS, or to alter appearance, cognitive function, athletic prowess and much more. Many feel strongly that it is far too early to attempt such therapy on human beings. Even Venter, often described as a 'maverick' biologist and famous for creating the first bacteria with completely synthetic chromosomes, urges extreme caution here: 'Just starting to do it on human embryos because we think we're curing disease is extremely dangerous.'[46]

CRISPR-Cas9 is more precise than any approach used before, but Venter still considers it a 'sloppy tool because ... CRISPRs have all kinds of off-target effects ... Instead of just changing the base pair that you want to change, they change a whole lot of others in the background.' Scientists are working hard to improve the accuracy of CRISPR-Cas9.[47] Other researchers have engineered variant forms that cannot cut DNA, but instead dial up, damp down, or wholly quench the action of genes in controllable ways, without changing the DNA itself. Eventually these subtler tools may be deemed more acceptable, since their effects will be less permanent. Venter remains unmoved: 'Even if CRISPRs were totally precise, there should be a little bit more public agreement before going ahead with human editing ... As you will recall, at the end of World War Two, after the Nazi atrocities, we decided not to do human experimentation.' The urge to engineer, fix and improve problems runs deep. As these techniques become more proven and familiar, it is inevitable that they will be brought to bear on human DNA, and perhaps this represents the logical conclusion of recent centuries of medical advance. Since ageing is programmed into our genetic make-up, some researchers think that gene-based therapies will eventually be the most effective way to subvert that pressure and evade mortality altogether.

Geneticist George Church works at Harvard's Wyss Institute for Biologically Inspired Engineering. Throughout our animated conversation, he stands at his laptop. He seems to be answering emails as we speak, but gives no outward sign of being distracted. He refuses to believe that decrepitude and death need to be our ultimate fate, saying, 'I don't buy it when somebody says something is impossible. I've banned the word in my lab.' Church has no time for wishful thinking and does not believe in 'this "fountain of youth" kind of thing, where there's going to be some source of water, or nutritionals, or small molecules that we'll just pop and live forever'. Instead, he approaches the problem of ageing as an engineer. 'I think that ageing is about multiple failures, all of them having the appearance of engineered planned obsolescence,' he continues, 'so you have to attack all those systems and you have to do it with powerful medicines, not things you find in plants that just happen to serendipitously solve your problems.' He agrees with Langer that 'the most powerful medicines are gene therapy' and describes forty-five experiments underway in his laboratory that are designed not just to arrest but also to *reverse* specific symptoms of ageing in laboratory animals, by manipulating the actions of genes. In another set of experiments, he is using gene-editing techniques to modify the genomes of pig organs, in order to make them safe for transplantation into humans. Eventually, Church thinks he will be able to make pig cells resistant to cancer, pathogens and the normal deterioration associated with ageing. 'Maybe once we see enough transplant successes with enhanced pig organs, there will be a path to executing similar procedures directly with human organs,' he says.

Once this still-notional programme of ageing prevention and reversal strategies is delivered, Church believes it will become a must-have treatment. 'The people who talk against it are essentially talking suicide, even though they frame it in a different way. It just is,' he argues. His confidence is extraordinary, but it is founded on a strong academic and commercial track record. He holds dozens of patents and has launched eleven successful companies.

Across the Charles River at MIT, I describe Church's bold plans to Langer. He clearly has great respect for Church's approach to discovery and innovation and he does not dismiss Church's claims, but instead injects a strong dose of pragmatism. 'There will be progress,' he says, but 'I guess the question for me is how long we will have to wait before these things happen. You can discover things or invent things fairly quickly, but moving them through patients – that's a long, slow process, and it's never been anything else.' Recalling the terrible effects of the thalidomide disaster during the 1960s, when a supposedly safe sedative caused the birth of at least 10,000 physically impaired babies, Langer underlines the need for exacting safety testing. Nevertheless, he does think that today's regulatory organisations are sometimes overly cautious. He also laments the 'incredibly litigious society [in the US] … Anybody can sue anybody here … that also engenders extreme caution.' Despite these obstacles, Langer shares Church's belief that eventually it will be possible to replace or rebuild whole organs and tissues, using re-engineered human and non-human animal cells, synthetic parts and electronics. 'I do believe that someday it will happen, but it may be many years from now,' he says.

The sheer complexity of the human body, combined with the generally conservative leadership of medical institutions, means that we are not on the verge of creating sub-species of enhanced, eternally youthful human beings. But, as in all the other activities described in this book, there are important trade-offs to consider. The technical question of whether a life *can* be extended must always be balanced against the wider issue of whether it *should* be. And, if the causes of natural death can eventually be eliminated, would the population grow to an unsustainable size, or would removing the imperative to reproduce cause it to dwindle? At the crux of these questions is the value that we place on a human life and how much we should pay to preserve it. Different individuals living in different communities, be they rich or poor, old or young, in the East or in the West, will arrive at different conclusions.

Engineers save far more lives than physicians, and this will continue to be the case as they expand public health interventions throughout the developing world and create the technologies needed to diagnose and cure illnesses ever earlier. Treatment will become increasingly targeted, but we still will not see wholly personalised medicine, since that would cost too much to develop and regulate. Innovative genome analysis and genetic manipulation techniques will prevent the agony of certain inherited genetic conditions; however, conventions must be put in place to limit the genetic manipulation of healthy humans, especially since techniques are now much more powerful than the eugenic practices advanced by the Nazi regime. Country-by-country regulation is also necessary, in order to avoid a global crisis caused by the abuse of antibiotics and the consequent creation of antimicrobial resistant infections. Average life expectancy will continue to rise and, as populations age and health care costs increase, there will be a shortage of physicians and other medical staff. These professionals will need all the assistance that robotics, artificial intelligence and advanced screening devices are able to provide. This new wave of medical automation will compensate for gaps in health care and, just as importantly, it will release the labour needed to care for an ageing population. Nothing expresses better the correct mission of medicine than Bob Langer's parting words to me: 'I just want to see us relieve suffering. I think that would be a good thing.'

10

Imagine

Unanswered questions give free rein to the imagination, which is probably why I find such intrigue in the archaeological remains of ancient civilisations or the mysteries surrounding the origins of artistic masterpieces. The search for such answers connects me with those who have pursued answers to questions about the beginning of time, the most elementary components of matter and the nature of human existence. Marvelling at the universe 'in the whole extent of its infinite grandeur', Immanuel Kant described, in 1750, how confronting the edge of our understanding can inspire us to action, when 'a vast field lies open to discoveries, and observations alone will give the key'. The quest for answers inspires us to build machines and systems that help us explore, experiment and evaluate the state of our world; even if satisfying answers elude us at first, we must never stop enquiring. The search helps to broaden our perspective, and the knowledge we glean along the way is potent fuel for the imagination, the driver of all progress.

ZOOMING IN
Our everyday experience of the world depends on the behaviour of invisible forces and particles that exist at the smallest scale. Engineered systems allow us to peer inwards and confront our ignorance about these fundamental processes, so I went to see how we look at very small things by using a very large marvel of engineering.

Just south of the main campus of Stanford University, a slender concrete tube traces a perfectly straight, two-mile-long line towards the Californian hills. One blistering hot summer's day, the air thick with the smell of sun-baked pine trees, I walked for a few hundred yards through this building, which contains the Stanford Linear Accelerator Center (SLAC), designed to reveal the workings of the sub-atomic world. Inside, devices called klystrons, placed every dozen or so yards, produce a loud buzzing noise. Creating synchronised pulses of microwave energy, they then convey them to a narrow copper tube, which is surrounded by strong electromagnets and buried thirty feet below the building. The energy from the klystrons and the force of the magnets accelerate electrons, which are fired into one end of the copper tube, until they are travelling at a velocity close to the speed of light.

In a series of experiments conducted at SLAC in the late 1960s, researchers collided beams of accelerated electrons with positively charged hydrogen atoms, called protons. By analysing the aftermath of these collisions, they were able to gather the first experimental

FIGURE 10.1 Looking at small things with a very large thing: SLAC, Stanford University, constructed in the 1960s.

evidence that confirmed the existence of quarks, abruptly changing our understanding of the nature of matter. For centuries, scientists had been stripping substances down, trying to find the universe's smallest, most elementary building blocks. For a time, the proton seemed to fit the bill – it was clearly a particle and had, until that point, seemed self-contained and indivisible. However, experiments at SLAC proved that such a neat solution was wrong. Protons can be divided into smaller particles – each is made from three quarks that, together with leptons, are a key component of the 'Standard Model' of particle physics. For the time being, quarks stand as genuine elementary particles that have so far resisted theoretical and experimental attempts to break them down into even smaller fundamental parts.

These days, the linear accelerator at SLAC is no longer the most powerful tool for analysing fundamental particles. Another extraordinary engineered structure, the Large Hadron Collider at CERN, near Geneva, superseded it when it opened in 2009. It was at CERN that a crucial missing piece of the Standard Model fell into place, when a new fundamental particle, the Higgs boson, which is thought to give all other particles mass,[1] was detected in 2012. In the meantime, SLAC has been repurposed to answer an entirely new set of scientific questions, having become the site of one of the world's most powerful X-ray lasers. This laser uses a long section of the original SLAC accelerator to produce coherent bursts of X-ray light, which are delivered for a few femtoseconds (each one being a millionth of a billionth of a second). The light from the Linac Coherent Light Source is billions of times brighter than any other light source that had been built before and, as the project's director Mike Dunne explains, 'when you get a factor one billion change in any parameter, you know that something pretty dramatic is going to happen'.

The properties of this light are perfect for studying 'ultra-small' objects and 'ultra-fast' processes. SLAC's laser can show how nature works, by capturing high-resolution images of the individual atoms in molecules as they shift and interact. Understanding how that happens could transform our world. For example, chemical processes, such as the expensive and inefficient Haber-Bosch process that converts nitrogen from the air into artificial fertiliser, could be re-engineered. New generations of catalysts could be designed by 'watching' processes occur at the atomic level, and that could also be done to aid the design of drug molecules for diseases such as Alzheimer's. Dunne goes on to describe how the observations made at SLAC are helping to design qubits for quantum computers. When capable quantum computers exist, they will be able to model the 'digital twins' of the physical phenomena observed by the laser, opening up ever more possibilities for simulation and engineering at the atomic scale.

Making this facility work has been an all-encompassing challenge. 'The essence of a facility like this,' as Dunne explains, 'is the confluence in time of all of these different state-of-the-art technologies – detectors, optics, computing, sample environments – and getting those integrated in a robust form.' For chief engineer Nicholas Kelez, the project has been 'as much of a social engineering challenge as a mechanical engineering challenge. People can be very elastic; you can push them, but as soon as you let off, they just go back to where you started.' Kelez and Dunne appear to have overcome much of that intrinsic inertia, and have found a way to make diverse teams of engineers, scientists and technicians work together in imaginative and productive ways. Engineering and scientific enquiry are symbiotic – each nurtures the other, which encourages both to grow. Just as the ability to engineer new scientific apparatus depends upon knowledge gathered by scientists, every branch of scientific enquiry leans heavily on engineering. Without effective tools, discovery slows. This is true of the science of the smallest things, like the atomic and molecular processes studied at SLAC, as well as the science of the very biggest things, including our attempts to understand the cosmos itself. It is also true of the biological processes which make up the brain and which work at a scale between these two extremes.

THE NEURAL FRONTIER: MAPPING AND MANIPULATING
THE BRAIN

Biology is the focus of the Crick Institute in London. Four floors beneath street level is the Institute's suite of electron microscopes, each of which balances on a hefty concrete pad, suspended on air springs to keep it completely stable. I watch as a scanning electron microscope focuses a beam of accelerated electrons on a small piece of brain tissue. As they hit the sample, secondary electrons are emitted from the surface. The microscope captures those electrons and analyses their properties to build up a high-resolution, three-dimensional image of the sample under scrutiny.[2] The investigators leading this study are attempting to reconstruct the connections that make up one of the brain's circuits. It is these circuits that allow us to sense our surroundings, experience emotions and consider the future. When they misfire or incorrectly rewire, brain disease and dysfunction can follow. Understanding the structure of these circuits and how they work is a formidable challenge; a single cubic millimetre of cerebral cortex, a piece about the size of a poppy seed, contains at least 50,000 neurons; each sprouts a thicket of fragile protrusions, called axons and dendrites, which permeate the brain, forming synaptic connections with approximately 6,000 other neurons. Together, there are at least 300 million synapses in every cubic millimetre of brain, making it one of the most complex structures imaginable.[3] As John Hennessy, one of the pioneers of computer microchip design, explains to me, 'we're not even close to having a machine of similar capability, not even close. You can take all the computers in the world, but one human brain has more synapses than all those connections put together.'

The electron microscope is a powerful way to map and understand the structure of brain circuits on a small scale. The experiment that I saw hoped to provide new insights into the way animals perceive smell. However, electron microscopy is an arduous way to map an entire brain – it would take a year of scanning to image the contents of a millimetre cube of human brain, at which rate the whole of the human cerebral cortex would take 500,000 years to map. Interpreting those images and constructing a map of a brain's neural connections cannot yet be done, since to do it would take more than the entire world's current computing power. And even if you succeeded in this challenge, all you would have is a static snapshot of a system that is actually in flux – your brain continually rearranges its physical structure as you learn and experience the world.

Despite all these difficulties, many scientists believe that we will at some point need detailed maps of the whole brain, in order to understand how it works and to fix it when it goes wrong. As the young MIT researcher Ed Boyden puts it, 'We've got to map the brain first, but we don't yet have a complete map of any brain, not even a worm's.'[4] To make these maps will require new technical approaches. Boyden thinks he has developed a powerful new technique that will have a big impact. His self-confidence comes from his proven ability to turn convention on its head; he knew that much of the challenge of studying the brain stemmed from the minute scale of its components, so resolved that things would be simpler if he enlarged the brain physically – and in 2015 he did exactly that.[5] When water was added to preserved brains seeded with pieces of the polymer used in baby nappies, they expanded by between one hundred and 1,000 times. Boyden can therefore see individual biomolecules, the building materials of neurons, easily and quickly, using conventional visible light microscopes. His plan is first to map the simple brains of worms and mice, after which he wants to attack the daunting complexity of the human brain. He thinks he can get the mouse's 100 million neurons and 100 billion synapses mapped within 'ten to fifteen years', but 'that may be too pessimistic. I like to under-promise and over-deliver,' he adds.

Describing his motivation, Boyden says that 'for me it's about the philosophy, it's about understanding the nature of existence. And as a by-product let's cure everything.' He believes that mapping the brain is the first step to curing all brain illnesses. That is increasingly important, since depression is the leading cause of illness and disability in the world,[6] anxiety disorders blight nearly as many lives, and brain degeneration and dementia are becoming more widespread as life expectancy is extended. He says, 'I find it hard to believe that if we map everything, see everything and control everything, we will have zero approaches for therapy'. Understanding the architecture of the brain is an important start, but if we want to cure or manage its malfunctions, we also need to understand its language, which is what Justin Sanchez is trying to do at DARPA.

Sanchez is making progress in his aim to make machines that can understand the codes that the brain uses to generate perception, thought and action. We learned in Chapter 4 how his developments are giving mobility and sensation back to paralysed people. DARPA-supported research has also shown that electrodes implanted in the brain can improve memory formation and recall. They are now attempting to reach further still, tackling illnesses that are caused by problems in much more complex brain circuits. 'Many of our military personnel come back from service with serious neuropsychiatric illnesses,' Sanchez says. 'So they get drugged up, or they lie out on the couch and talk to their therapists. Maybe those [traditional therapies] work sometimes,' he continues, 'but they're not as precise as we want them to be and they're not as effective as we want them to be.' Sanchez freely admits that, when it comes to conditions like depression, anxiety, post-traumatic stress disorder and addiction, ignorance looms large. 'Where are they in the brain? Where are those neurons that are either firing the right way or the wrong way [to cause these conditions]?'

By working with patients who are having brain surgery to treat severe epilepsy, Sanchez's team has pinpointed the neural networks that misfire during bouts of extreme anxiety[7] and learned how to correct them by using implanted electrodes to stimulate specific groups of neurons. 'We turn on the system and then patients can respond almost immediately, saying "My anxiety level is going down. I feel kind of cool, calm and collected,"' says Sanchez. He hopes that this kind of short-term fix for anxiety will help sufferers to 'unlearn' distressing habits, so as to result in more lasting cures. 'The brain is a plastic organ, meaning that it is continuously changing,' Sanchez explains. 'If we give it the right feedback, maybe it can learn to operate in the right state again.' Devices that electrically interface with the brain have a huge potential to help people with neurological damage and dysfunction. In a domain where conventional pharmaceuticals and therapies often have limited success, these technologies may eventually prove to be transformative. In some situations they already are – digital cochlear and retinal implants convey useful and life-changing sensory information to the brains of hundreds of thousands of deaf and blind people. The resolution of sound and visual images that these bionic implants provide is still limited, but will improve. Stimulating electrodes, placed deep inside the brain, can also help many people with Parkinson's disease, epilepsy and even severe obsessive-compulsive disorder to live less disrupted lives.

It takes little imagination to extrapolate beyond Sanchez's clinic to a future where brain-to-machine interfaces are available to anyone who wants to tune up their cognitive faculties. Some hope that these neural interfaces will be the ultimate augmentation for brains that seem frustratingly limited in their capacity and hard to control. For example, I often find myself thinking how convenient it would be if my brain could absorb and digest whole books and articles, without me having to read them word by word. Or what if we could tap into the rich seams of knowledge and computing power on the Internet by thought alone? Imagine being able willfully to tune out intrusive worries or improve your ability to think creatively. And, going even further, what about transcending the limitations of language to communicate precisely what you are thinking with colleagues and loved ones?

Anders Sandberg, of Oxford University's Future of Humanity Institute, thinks cognitive enhancement, whether mediated by a neural device, a new drug or a new therapy, 'is an amazingly important thing that people tend to overlook. Our society, our economy, is built on knowledge and skills, so if you find ways of amplifying or multiplying them, you are going to really boost our society.' These ideas make many people deeply uncomfortable; more than two-thirds of American adults said they were worried by the prospect of brain implants that would enhance cognitive abilities.[8] There are, of course, many possible pitfalls. Granting technology direct access to our thought processes could make us vulnerable to commercial or political coercion, or worse. Nevertheless, as Sanchez says, 'Neural interfaces are opening up conversations about the possibilities of what we might do with our brains.' If we have these conversations as early as possible and make them broad in scope but always grounded in evidence, we have the best chance of biasing this technology to the good. Given the power to control your state of mind and emotional balance, would you opt for a constant state of bliss equivalent to the effects of the fictional drug 'soma' in Aldous Huxley's *Brave New World*? Or, as I suspect is more likely, would you opt for a more rounded inner life of expansive highs and introspective lows? Sanchez and Sandberg believe that these capabilities could be just a few steps away.

Boyden thinks there is still a long way to go before we will be able to enhance mental capabilities in a routine or reliable way. 'The honest answer is we don't know enough about the brain to know exactly what we can augment,' he says. Sanchez concurs and notes that 'it is easier to get large improvements when people have an impairment of their brain. It's harder to get improvements when people are already operating at their natural limits. We don't really know why.' Boyden identifies another challenge: 'I think that everything in the brain is so entangled. There's been no selective boosting of one thing without some kind of side effect,' he says. He points to the many thousands of people who have received implants since 1986 for deep brain stimulation and says that 'a lot of them develop strange behaviours. They can become very impulsive, making bad decisions and losing their money, for example.' As a result of these problems, neural interface technology is not poised to take over. There is still neither a direct brain–machine interface on the market nor any prospect of neurosurgeons operating on healthy brains to install one. If you are eager to upload your consciousness to the cloud, I am afraid you have quite a wait ahead of you.

THE PROMISE AND PERIL OF ARTIFICIAL INTELLIGENCE

One of the drivers behind our attempts to map the brain is the desire to see if we can build a silicon-based replica with human-like intelligence. And if we could do that, could we then build the next generation of computer systems that think, dream and reason for and with us? Artificial intelligence seems to be everywhere. Magazines and newspaper articles promote it endlessly, raising expectation and fear in roughly equal measure. As we saw in Chapter 3, some forms of AI are becoming ubiquitous: for example, algorithms execute huge volumes of trading on our financial markets, self-driving automobiles are beginning to navigate city streets and our smartphones are translating from one language to another. These systems are sometimes faster and more perceptive than we humans are. But that is only true for the tasks for which the systems have been designed, and that is something that designers want to change. They want to move on from today's world of 'weak' or 'narrow' AI, to create 'strong' or 'full' AI, or Artificial General Intelligence (AGI). This may become a reality because of the powerful computing machines, vast data storage and communications capabilities which are beginning to make our brains look puny.[9] If these devices could be provided with more flexibly intelligent algorithms, the opportunity would be huge. AGI could, its proponents say, work for us diligently, around the clock and, drawing on all available data, could suggest solutions for problems such as providing pre-emptive health care, avoiding stock market crashes or preventing geopolitical clashes. Google's DeepMind, a company focused on the development of AGI, has an immodest ambition to 'solve intelligence'. 'If we're successful,' their mission statement reads, 'we believe this will be one of the most important and widely beneficial scientific advances ever made.'

Since the early days of AI, imagination has outpaced what is possible or even probable. In 1965 an imaginative mathematician called Irving Good, who had been a colleague of Alan Turing at Bletchley Park,[10] predicted the eventual creation of an 'ultra-intelligent machine' that can far surpass all the intellectual activities of any man, however clever'.[11] Such a machine would be able to turn its vast intellect to improving itself – each tweak would improve its ability to enhance its own powers, leading to a rapidly accelerating positive feedback loop. 'There would then unquestionably be an "intelligence explosion",' Good wrote, 'and the intelligence of man would be left far behind.' He suggested that 'the first ultra-intelligent machine' could

FIGURE 10.2 The all-knowing, all-understanding *Brazen Head*, purportedly created by thirteenth-century philosopher-friar Roger (1748).

be 'the last invention that man need ever make'. This led to the idea of the so-called 'technological singularity' proposed by Ray Kurzweil, which argues that the arrival of ultra-intelligent computers will be a critical turning point in our history, beyond which there will be an eruption of technological and intellectual prowess that will alter every facet of existence. Good added an important qualification to the 'last invention': the idea that we would be able to harvest its benefits 'provided that the machine is docile enough to tell us how to keep it under control'. In Chapter 3 I discussed the challenges posed by 'black box' algorithms, or those that make biased and unaccountable decisions. Fears about the advent of malign, powerful, man-made intelligent machines are reinforced by a lot of fiction – Mary Shelley's *Frankenstein* and the *Terminator* film series, for example. But if AI does eventually prove to be our downfall, it is unlikely to be at the hands of human-shaped forms like these, with recognisably human motivations such as aggression or retribution. I agree with Oxford

University philosopher Nick Bostrom, who believes that the gravest risks from AGI do not come from a decision to turn against humankind but rather from a dogged pursuit of set objectives at the expense of everything else.[12] Bostrom says that confronted by 'the prospect of an intelligence explosion, we humans are like small children playing with a bomb. Such is the mismatch between the power of our plaything and the immaturity of our conduct.' Stuart Russell summarises what he sees as the core of this problem: 'If you, for example, say, "I want everything I touch to turn to gold," then that's exactly what you're going to get and then you'll regret it. If the machine is more intelligent than you, you won't be able to necessarily interfere or change its course of action.' If computers do become extremely intelligent, there is no reason to expect them to share any capability that people would recognise as justice or compassion.

The late Stephen Hawking told the BBC in 2014 that 'the development of full artificial intelligence could spell the end of the human race,'[13] and since then Bill Gates and Elon Musk have delivered similarly high-profile warnings. Recently Musk revealed that part of his motivation for investing in neural implant technology is his belief that humanity's best hope for controlling ultra-intelligent AI will be to couple with it directly. Only if ultra intelligence is an integral part of us, he argues, will we be able to ensure that it shares our values and align its moral compass with ours. Sandberg, however, is not convinced that this logic stands up to scrutiny: 'I'm already merged with a machine, because I have a smartphone. Part of my mind already resides in the cloud. But if an artificial intelligence converts the universe into paperclips, that doesn't mean I was on the winning side.' He also believes that 'you don't necessarily have to postulate exponential increases [in intelligence] to get into trouble', but imagines much more incremental improvements, which would eventually mean that 'someday you'll have a world where there's a lot of smarts for whoever pays for it'. Sandberg thinks it would only be a matter of time before somebody asked a machine, 'How do I take over the world?'

BRINGING AI BACK DOWN TO EARTH

Sandberg's colleague at Oxford the philosopher Luciano Floridi writes, 'True AI is not logically impossible, but it is utterly implausible.'[14] I hear this view from almost all the serious AI researchers that I meet. 'We are several algorithmic breakthroughs away from having anything that you would recognise as general-purpose intelligence,' says Russell. Tencent's Tong Zhang agrees: 'If you want general AI, there certainly are a lot of obstacles you need to overcome.' 'I just don't see any practical drivers in the near future for a cross-sectional general superintelligence,' adds MIT roboticist Cynthia Breazeal, who believes that 'Right now, commercial transformation is funded and driven by entities seeking directed solutions, delivered by highly crafted, special-purpose systems.' Mark James of Beyond Limits also doubts that anyone is really on track to develop true AGI, saying that 'for the AI field to truly progress to the point of having a really human-like thinking machine, we need to rethink the problem from square one'.

I think James is right – after all, how can we engineer something that we cannot even define? 'I'm a singularity sceptic,' says computer scientist Steve Furber, who makes the point that 'we've never really stopped to work out what natural human intelligence is, so we don't really know what it is we're trying to imitate in our machines'. Furber goes on to note that there are many types of intelligence, including emotional, musical, sporting and mathematical intelligences, saying, 'I don't think intelligence is a single physical parameter.' Zoubin Ghahramani concurs: 'I actually don't think there is such a thing as general intelligence,' he tells me. And if there is no such thing as a general intelligence, there is no hope of building one, either from synthetic or biological parts. Ghahramani goes further still. He thinks 'our view of intelligence is "pre-Copernican". We have a certain hubris where the things we're terrible at – we just say "That's not really intelligence." The things we're good at we say are what intelligence really is.'

Leading AI researcher Michael Jordan, from the University of California, Berkeley, points out that civil engineering did not develop by attempting to create artificial bricklayers or carpenters and that chemical engineering did not stem from the creation of an artificial chemist,[15] so why should anyone believe that most progress in the engineering of information should come from attempting to build an artificial brain? Instead, engineers should direct their imagination towards building computer systems that think in ways that we cannot: that grapple with uncertainty, calculate risk by considering thousands or millions of different variables and integrate vast quantities of poorly structured data from many different sources. Jordan asks that also we move away from the acronym 'AI', which has become associated with emulation of human intelligence, and focus instead on 'IA', meaning 'intelligence augmentation', and 'II', 'intelligent infrastructure', which he hopes will become 'a web of computation, data and physical entities that exists to make human environments more supportive, interesting and safe'.

None of this is to take away from the power of increasingly adaptable AI algorithms, or to ignore the risks that they could one day pose through unanticipated side effects or malign applications. If we believe that a machine with generalised human-like intelligence is impossible, many concerns about AI evaporate; there is no need to write any rigidly defined moral code or value system into the workings of AI systems. Instead, our aim should be to make them controllable and highly responsive to our needs. Many first-rate researchers and thinkers, including Russell and Sandberg, are devoting a great deal of time and energy to pre-empting problems associated with AI before they arise. Russell thinks the key to making AI systems both safer and more powerful is in making their aims inherently unclear or, in computer science terminology, introducing uncertainty into their objectives. As he says, 'I think we actually have to rebuild AI from its foundation upwards. The foundation that has been established is that of the rational [human-like] agent in optimisation of objectives. That's just a special case.' Russell and his team are developing algorithms that will actively seek to learn from people about what they want to achieve and which values are important to them. He describes how such a system can provide some protection, 'because you can show that a machine that is uncertain about its objective is willing, for example, to be switched off'.

Work like this is important, particularly because Russell and his collaborators are not simply flagging up ill-defined risks, but also proposing concrete solutions and safeguards. This is what Google's AI expert Fei-Fei Li meant when she said to me, 'it's not healthy to just preach a kind of dystopia [about AI]. It's much more responsible to preach a thoughtful message.' If the only message presented about the great leaps in physics during the early twentieth century had been dire warnings of imminent nuclear Armageddon, we would not now have all the amazing discoveries that have stemmed from our understanding of atomic structures and quantum mechanics. The world witnessed the dreadful potential of that technology when atomic bombs were dropped on Hiroshima and Nagasaki, but since then, no nuclear weapon has been detonated in war and the world's nuclear power sector has an excellent safety record. As computer scientist and investor Kai-Fu Lee writes, 'I worry about the cold war analogy ... AI is mostly an enabler, it's not a nuclear weapon.' And, referring to the fact that many AI researchers are eager to share their advances with the community, he thinks that 'it's probably the best worldwide collaboration anyone has ever seen'.[16]

The risks associated with AI must be kept in perspective and responded to with constructive action and regulation, rather than hand-wringing and alarmism. Tong Zhang points out that in China a more pragmatic route is being pursued: 'You have to think how to do the more positive things with technology, rather than thinking too much about the negative things. In China you can get a lot of data ... and people are more used to change. New applications emerge faster in China.' A Chinese robot recently passed a written medical qualification exam, giving a foretaste of things to come.[17] Between the extraordinarily optimistic and the terrifyingly pessimistic lies a more realistic future for AI. Long before they achieve anything even remotely resembling ultra-intelligence, computers will continue to change how we live and think in ways that are both far-reaching and hard to predict. As our computers become smarter, people will also get smarter and more capable. We will need the processing power and distinctly non-human insights generated by machines to take on our most pressing global challenges and to seek answers to the deepest questions. AI is already helping make sense of the cascades of information coming from SLAC, CERN and the other large-scale scientific experiments around the world. It will help us understand the maps of the brain and the neural recordings we make as they become richer and more detailed. And we will definitely need machine learning to comprehend what we see as we zoom out and confront the vast timescales and immense forces that make up the universe itself.

ZOOMING OUT

Plans are in place for the launch of a rocket that will take off from a site near the equator in French Guiana in the spring of 2020, before puncturing Earth's atmosphere and delivering its payload into orbit. Over the subsequent thirty days, this payload will, like a butterfly slowly unfurling from its chrysalis, become an elegant spacecraft. If all goes to plan, this new interplanetary object will bring completely new views of our cosmos. Even if it does not work, the mere idea that such an audacious feat of engineering is possible should fill us with inspiration. Under a blue sky in early April, I visited NASA's Goddard Space Flight Center to see the James Webb Space Telescope as it was taking shape. The grey, down-at-heel corridors of empty offices and obsolete printers made the place feel more like a mid-twentieth-century public institution than a place where people were about to explore the outside edge of astronomical knowledge and the limits of engineering possibility.

'When they first started teaching me how to make spacecraft, thirty-seven years ago, they told me "don't deploy anything",' Mike Menzel, Mission Systems Engineer for the world's newest and most powerful space telescope, tells me from his high-backed office chair.

'For what will be my final mission, I'm breaking all the rules.' The telescope's final destination is the second Lagrange point, nearly a million miles from Earth.[18] After the month-long journey there, the bulky satellite in the huge room below me will undo the 140 precise folds that will have squeezed it into the rocket that will launch it into space. A sunshield made from five individual layers of foil-like Kapton film will unroll, creating a diamond-shaped parasol with the same area as a tennis court. 'If you think that happens in zero gravity without a lot of mechanisms to control it, you're crazy,' offers Menzel. Protected by the sunshield is the telescope's main mirror; sixteen burnished hexagons, made from gold-plated beryllium, intersect to form a flawless and convex reflective surface, six and a half metres in diameter.

FIGURE 10.3 Mirror, mirror into space, will it unfold in outer space? The James Webb Space Telescope under construction in 2017.

Soon after the Hubble Space Telescope had been launched in 1990, it became clear that its main mirror, one of the smoothest optical surfaces ever produced, was fractionally misshapen, which meant that the images it produced were severely compromised. NASA engineers corrected the flaw by visiting Hubble, which orbits 350 miles above the Earth, and installing a component that reversed the aberration, in effect giving Hubble spectacles. The James Webb Space Telescope is Hubble's successor, and will be the world's most powerful and sensitive telescope; once it is in place, it will be so far from Earth that a return visit is not an option, which means that Menzel's team have just one chance to get it right and make good on the $10 billion investment that the project has received. Menzel describes the fast-approaching first days of deployment as his 'fourteen days of terror'.

There is a great deal that could go wrong, and the chance of failure escalates dramatically as the deployment steps become more numerous and complicated. But Menzel, with a neatly cropped beard and sharp eyes behind round, wire-framed glasses, radiates calm, confident authority. To build the telescope, he has combined a meticulous approach, focusing on small details and relentless testing, with a wider overview of the system as a whole. I find it easy to believe that this fearsomely ambitious engineering project will roll out as planned. Some of the infrared photons that James Webb's mirror will gather and focus onto its sensitive detectors will have travelled for more than 13 billion years, from their source in the fiery cores of our universe's very first stars;[19] nobody has yet been able to observe these ancient celestial bodies. Astronomers cannot predict with any certainty what they will learn from their new observatory in space – Menzel reminds me that 'today we live in a universe where we don't know 95 per cent of its content'. The forms of matter and energy with which we are most familiar, the forces and elements that engineers and inventors manipulate every day, are in fact relatively rare in the cosmos; the bulk of the universe's matter is poorly understood dark matter and most of its energy is the even-more-mysterious 'dark energy'. Menzel's new telescope may confirm or wholly revise our present cosmological hypotheses, and he thinks it has the potential to deliver the first strong evidence of life on other planets – its results could be just as revolutionary as the heliocentric theory first devised by Nicolaus Copernicus and reinforced by Galileo Galilei five centuries ago. The first telescope destabilised humankind's view of itself, throwing the Christian church into disarray and threatening to upend society entirely. If there was no heaven, might there be no God? And if there was no God and no heaven, what could possibly maintain the world order?

SEEING THE BIGGER PICTURE

The James Webb Space Telescope promises to deliver the ultimate overview of the universe. And as Edgar Mitchell, a member of the Apollo 14 mission and the sixth person to walk on the surface of the moon, learned, seeing Earth from space can be a powerful experience. He described the feeling looking back at it as an 'explosion of awareness' and an 'overwhelming sense of oneness and connectedness ... accompanied by an ecstasy ... an epiphany'.

That profound sense of awe experienced by many astronauts is called the overview effect;[20] seeing the Earth from afar may be the greatest demonstration that we live in a vast and highly interconnected system. The same is true for many things – in all our activities, from science and engineering to politics, business and art, we should forget the overview at our peril. While many of the great advances of the twentieth century relied upon a tight focus, reductionist thinking and specialisation, the greatest challenges of the twenty-first century can be solved only by the consolidation of specialisations. This is very evident at the Crick Institute, where the Director, Paul Nurse, criticises the fractured relationship between science and engineering. Nurse, with tousled white hair, an open, direct manner and a ready smile, gets straight to the point. 'I hate these words: pure science, basic science, fundamental science. I never use any of them,' he says emphatically. 'I use the phrase "discovery science". The research spectrum goes from thinking about things that are not obviously

FIGURE 10.4 10 billion light-years away, but not out of sight: IDCS 1426, one of the earliest clusters of galaxies ever captured by Hubble.

useful to things that are very useful,' he continues. 'There's discovery throughout all of it.'

Nurse is right – the division between what others call basic science and engineering is artificial and counter-productive. 'Look back at the origins of modern science in the 1660s,' he says, the pitch of his voice rising with incredulity. 'This distinction was meaningless, completely meaningless.' He has no time for the arrogance of scientists who claim to be 'pure intellects' and refuse to be 'sullied with dirty commerce'. To be engaged in the process of discovery is to be alive to possibility. The routes between the laboratory bench, the factory floor and the shop shelf are many, varied and hard to predict. 'Look at the second law of thermodynamics,' Nurse urges, to illustrate his point. 'It was developed by people investigating steam engines during the

Industrial Revolution; rather than by pure scientists. The principles of thermodynamics were applied in a machine long before they were contained in a theory. There is not and never has been an obvious or linear conveyor belt that takes fundamental research and translates it into useful products. 'Ecosystem may have been overused as a metaphor; Nurse explains, but 'it is definitely a complex network.'

The research going on in Nurse's lab, the continuation of the work that earned him the 2001 Nobel Prize in Physiology or Medicine, is a case in point. For over four decades, Nurse and his colleagues have been deciphering the inner workings of single-celled fungi called fission yeast. The compulsion to explore and to imagine how things actually work delivered a stream of discoveries that formed the motivation for understanding and then tackling cancer in human cells. Nurse explains that this application was not the primary reason for his research, but that he always hoped that his discoveries would eventually prove useful. 'I have never sneered at that stuff; he emphasises.

Nurse describes the unhelpful schism that exists between science, engineering and commerce, but the rifts spread even further and deeper than that. 'We've seen walls grow up. Really, it's completely counter-productive; he tells me. He is clear that our core mission should be 'understanding the world and making use of that knowledge to improve the state of humankind; which he insists should be a 'joined up activity. We break that at our peril.' In his 1959 lecture, C. P. Snow pointed to a fundamental division of society into what he called 'The Two Cultures'. On the one hand was 'science; defined in its broadest sense, and on the other, 'the arts'. Snow made an impassioned plea for better bridges to be built between the two – his diagnosis was accurate, and too little has since been done to close the division. But the reality is that each of his two cultures is further divided into sub-cultures that, whether actively or through passive neglect, continue to erect barriers between one another. Some degree of specialisation is still necessary, of course. From the innards of a smartphone to the tangles of neurons in the human brain, no single person can fully understand the complex systems that we seek to build and manipulate. But to take on the sprawling, fast-changing problems of our age, we need people and organisations that can take an overview, understand the intricacies of the human condition and defend positive values. In short, we need to pool our intellects and use our imagination collectively.

I visit the designer Thomas Heatherwick at his studio, which is tucked away modestly behind a budget hotel in the London district of King's Cross. He is no ordinary designer, and his work and thinking defy easy categorisation. I put it to him that he is part-designer, part-curator, part-architect and part-engineer. 'I'm most comfortable describing myself as a designer, or perhaps "an expert at not being an expert",' he replies.

Heatherwick's studio embodies his multi-disciplinary approach. Two hundred people work here and as well as banks of computers,

FIGURE 10.5 Enjoy the view: Earthrise from the moon (by NASA, 2015).

all equipped with the best computer-aided design software, there are also well-equipped wood and metal workshops, 3D printers and lab-like spaces in which employees can tinker and think. For Heatherwick, making and thinking are inseparable. We sit on bright green plastic chairs to talk at a round glass table at the centre of the studio. We are surrounded by shelves populated by arcane curiosities, which provide more fuel for the imagination – a Chinese model cat waves its paw incessantly, and next to it is a carved coconut husk, a cowboy boot and a model of the crystalline structure of salt. Reflecting his collaborative ethos, Heatherwick scrupulously describes his relationship to his work in the plural: 'we imagine'. From this studio have emerged a succession of ingenious, often playful ideas made concrete. In 2010, Heatherwick created his 'Seed Cathedral' for the UK pavilion at the Shanghai World Expo. A ten-by-fifteen-metre box bristling with 66,000 transparent acrylic 'hairs' that moved with the wind and brought light into the structure. A few miles to the west of his studio, along the Regent's Canal, is a footbridge he designed that rolls up into a tight ball, like a caterpillar, when a boat needs to pass.

This sense of a free-ranging imagination, in combination with rigorous problem-solving, is a key feature of all of the best and most ambitious engineering projects. There is a playfulness in the notion of sending one of the most sophisticated sensing devices ever built, which unfolds like an origami flower, a million miles out into space, but there is also a steely rigour. Engineering encourages dreams to form and then casts them into reality. This happens at the bustling intersection between discovery, design and commerce, which is a rich and fulfilling place where hand and head work in harmony. But in today's world, too few people think of engineering as a creative enterprise. Heatherwick believes that engineering has allowed itself to become a 'silo' in which engineers simply 'crunch the numbers and make it work for the people who do beauty and human experience'. He is convinced that the connection between design and engineering in architecture has been broken, to the detriment of both disciplines. He is clear that 'engineering is about the human experience' and

FIGURE 10.6 A playground for imagining, designing and making. Thomas Heatherwick's studio, London.

believes that this 'might be an exciting moment to ignore the professional separations'.

Harvard biologist George Church has seen this tendency play out in his own field. 'Almost all major scientific progress, almost all disruptive, transformative steps forward are based on technology,' he says. Yet the people who develop these technologies often end up running facilities 'where they are literally servants' to other scientists. This distinction, which elevates the esteemed work of the head that thinks over the disparaged work of the hand that makes, has been played out time and again. The late Professor Calestous Juma indignantly described what he saw as the 'very systematic appropriation of credit from engineers and reallocation of it to scientists. People don't see the difference between technicians and inventors.' These divisions and false hierarchies are, of course, nonsensical; as Heatherwick puts it, 'what a limited role it must be for someone who can dream but who cannot make those dreams a reality. If we're all stuck in our silos, how is progress going to really happen? Because it's the insights across it all, the connections, that ensure we don't lose control of the world. We have to have people who can glue together all the specialists. That's why my studio is made up of all sorts of people with different backgrounds and training: makers, designers, architects, landscape designers, interior designers and project managers. We all work together.'

WHY WE NEED PATIENCE

Decades of exponential improvements in computing power have created unrealistic expectations about the rate of engineering advance that we should be making. The information revolution has changed our world profoundly, but it is only one facet of engineering innovation. Vaclav Smil describes the problems of over-extrapolation from microchip fabrication as 'Moore's curse.' Similarly, most of the experts I met described how their innovations diffuse into our lives at a much slower rate than we realise. 'There's a big difference between software-enabled products like Facebook and material technologies,' explains Professor Richard Friend of Cambridge University. Developing new hardware usually takes longer than software. He gives the example of the two decades that it took to commercialise the organic LEDs that now light up the latest smartphones and television screens. Computer scientist Andy Hopper, meanwhile, expresses his frustration at the slow introduction of smart sensor technology into mainstream manufacturing. 'People talk about exponential technologies … Hogwash! Hogwash!' he says. 'That may be true in consumer electronics, where people have an appetite for trying new things,' he adds, but 'that's rather different from taking anything of that sort and pushing it to the manufacturing sector.'

In medicine, robotic surgery pioneer Ara Darzi describes the regulatory complexities and the conservatism of the medical establishment which mean that 'sadly it takes ten to fifteen years' for new medical engineering to reach the point where it is in use. And, despite all the excited publicity about self-driving automobiles, Professor Daniela Rus, Director of the Computer Science and Artificial Intelligence Laboratory at MIT, explains that they will not rush into our lives, sweeping away all in their path. She points out that full autonomy is already here but only in constrained environments, at low speeds, and on private roads, which is known as 'Level 4' autonomy. Full integration, or 'Level 5' autonomy, will come eventually, but we are still a long way from being able to build these vehicles, since they must work safely and dependably on any road, at any time, and in any weather. This means that society still has 'a long time to think about what we need to do, in order to help with the transition and prepare people', says Rus. Zoubin Ghahramani points out that we have had machines that are faster and stronger than ourselves for centuries, but they have never triggered runaway advances. 'The fastest airplane goes a little faster than it did a few years ago ... but it doesn't go at the speed of light,' he says. Why should artificial intelligence be any different? We must be patient, because the reality is that most engineering advances arrive and change our world at a steady, incremental rate. There is solace in this more sober message; society has time to reflect, to minimise the risk of unexpected damage and to design effective regulation.

IMAGINATION COMES WITH GREAT RESPONSIBILITIES

Speaking at Harvard University, the author J. K. Rowling defined imagination as the 'uniquely human capacity to envision that which is not, and therefore the fount of all invention and innovation'.[21] She is quite correct in this assessment, because all novelty stops without the power to imagine and, moreover, putting yourself in someone else's shoes is the ultimate act of imagination. While the mathematics that underpins technological progress is, of course, devoid of moral bias, the interpretation of those numbers never is. Even when we have the best of intentions, we build our human biases and flaws into the fabric of our creations; when we are immersed in the technical detail of making things work, it can be easy to forget that the devices and services we introduce into the world will have wide-ranging and unpredictable effects on the lives of real people. For this reason, the argument that just because something *can* be done, it *should* be done, is indefensible.

In 1945 Vannevar Bush, an engineer who had a great impact on the innovation policy of post-war America, expressed the need to nurture engineers who could think both broadly and deeply. 'It is unfortunate when a brilliant and creative mind insists on living in a modern monastic cell,' he wrote. 'One most unfortunate product,' he continued, 'is the type of engineer who does not realise that, in order to apply the fruits of science for the benefit of mankind, he must not only grasp the principles of science, but must also know the needs and aspirations, the possibilities and frailties, of those he would serve.' In the seven decades since Bush made this plea for engineers to expand their world views and connect more deeply with their societies, too few of them have heeded his words. Bush's message is more urgent now than ever before.

As Thomas Heatherwick's work demonstrates so eloquently, part of the impact that engineered structures have on us is influenced by the aesthetic response that they provoke; for any designed object, how you feel is part of its function. For this reason, engineering and design are inseparable. Jony Ive, Chief Design Officer of Apple, expresses a similar sentiment. He is clear that design, beauty and engineering are inseparable: 'Mechanised objects created a disjoint between function and form, to the point where, in the smartphone today, there are an extraordinarily large number of functions with no intrinsic form.'

Both Heatherwick and Ive are less concerned with how things are now than with how things could be in the future – it is unfortunate that we so often neglect to ask technically minded makers to consider such abstract qualities as beauty and meaning. Engineers are capable of extraordinary technological feats; the more speculative technologies that we have discussed in this chapter and elsewhere in the book, including neural implants, robotics, nanotechnology, synthetic biology and advanced AI, may all mature to provide us with radical new capabilities. However, all this work will result in meaningful progress only if engineers can use their imaginations and come to understand the strengths, weaknesses and desires of the people who will use these new systems.

THE DREAM

I recently visited the Chajnantor Plateau, in the middle of Chile's Atacama Desert. It is one of the driest and most starkly beautiful places in the world. At an altitude of more than 5,000 metres, the cold air is thin and the sky is an endless expanse of the deepest blue. This high plateau is also home to an array of sixty-six flawless, concave dishes that receive radio waves from the far recesses of the universe. Each twelve metres wide, they rotate in perfect synchrony and together function as a huge telescope, the Atacama Large Millimeter Array, which gazes out into the universe and back into the depths of cosmological time. The pristine natural environment, the impeccable engineering of the telescope's components and an appreciation of the deep questions that they had been built to answer filled me with a profound sense of awe.

As far as we know, nothing out there in the vast unknown cares about what happens to us. Indeed, the odds seem stacked against us; our ancestors emerged into a remorseless world and were exposed to the onslaught of the elements, the constant menace of predators and disease and the selfish lapses of their fellow humans. Defying all probability, we have applied our imagination and channelled our ingenuity to build the vibrant civilisations of today.

In the beginning this was a solitary endeavour, as self-conscious apes gradually started to build tools and organise their worlds. The process of civilisation may have started slowly, but over the millennia it has gathered momentum, and each generation of inventor has learned from what was done before. The Scientific Revolution of the sixteenth and seventeenth centuries gave more focus and rigour to the engineering process, as our collective intelligence began to grow more quickly. Activity reached fever pitch during the Industrial Revolution and has accelerated ever since. Over time, we have enormously increased our capabilities and expanded our horizons; engineering is now a huge and highly distributed activity, and no other sector of human endeavour can avoid its influence. It shapes our economies, our democracies,

FIGURE 10.7 'Hello, is anyone out there?' The Atacama Large Millimetre Array spots things millions of light years away on the Chajnantor Plateau in Chile.

our philosophies, the creations of our artists and the discoveries made by our scientists. Without advances in engineering, all these sectors would ossify and our world would seem to contract. We need to always be alert to people who may wish to apply the great powers afforded them by technology to selfish or destructive ends, and should be willing to restrain them if necessary. However, the sum total of useful knowledge can only grow. Provided that the forces of ignorance are kept in check, engineering will continue its work: solving problems, building civilisation and delivering dreams.

Staring up into the blue infinity of the desert sky, it seems obvious that we are not at the centre of this universe. Nicolaus Copernicus extinguished that idea five centuries ago, and since then, a series of great leaps in human thought have pushed us to renounce the idea that our species occupies a position of special privilege. Thanks to Charles Darwin, we know that we are not exceptional in the biological world. Then, in the nineteenth century, Sigmund Freud destroyed the notion that we are uniquely rational agents. Now, thanks to Ada Lovelace, Alan Turing and the subsequent generations of computer scientists, it is clear that we are neither the only, nor the most capable, intelligent agents on this planet.[22] To believe otherwise is exceptionalism. However, if I was forced to choose between a world governed by our familiar brand of human intelligence, albeit riddled as it is with biases, flaws and imperfections, or the cleaner, clinical intelligence of machines, I would have no hesitation in making my choice. For we have the special gift of imagination that allows us to dream. And all progress starts with a dream: a vision of a better world.

ACKNOWLEDGEMENTS

While I was writing this book, I was very mindful of the enormous debt I owe to my colleagues who are engineers, economists, business people, commentators and from many other walks of life. They have formed my view of the world. Their achievements are humbling and their insights are instructive. A few years ago I read Bill McKibben's book *The Age of Missing Information*, in which he challenges the view that we are better informed than any previous generation. In 1993 it seemed contentious; today, we might be sure that we have a lot more information, but are we really better informed?

In the face of that challenge, I decided to write this book. To give myself a greater chance of succeeding, I decided to interview over one hundred extraordinarily expert people, most of whom, in one way or another, I had previously met or collaborated with. Their insights form the platform upon which this book is built, and their biographies are all included in an Appendix. I would like to thank each of them for the time and care they took over these interviews; I hope they found them enjoyable and approve of the way in which their views have been used.

In addition, I asked several of my friends to read the whole or part of the manuscript in various stages of development. I owe a huge debt of gratitude to Lionel Barber, Nick Butler, Diane Coyle, David Halpern, Tony Fadell, Jeremy Farrar, Peter Hennessy, Andy Hopper, Donna Leon, Paul Nurse, John Thacker and Ed Williams. Mark Dominik and Matthew Powell from my office also gave valuable comments, for which I thank them.

I also want to thank: Nick Humphrey for a splendid job of editing the text; Alexis Kirschbaum, my editor at Bloomsbury, for guiding the development of the book; my literary agent Gordon Wise at Curtis Brown for his insightful comments; James Brimstead, Alice

Holt, Toby Gunton and Emma Zadravetz of Edelman for their extraordinary work in taking raw data and making understandable graphic material; and my partner Nghi Nguyen, for an insightful set of captions and his usual general editorial support.

Finally, I would like to thank my research colleague Ben Martynoga, for dedicating almost two years of his life to this project. He has been completely indispensable, and without him this book would never have seen the light of day.

John Browne
Venice, September 2018

BIOGRAPHIES OF INTERVIEWEES

AJ Abdallat is Chief Executive Officer of the artificial intelligence company Beyond Limits, which he founded in 2014 to commercialise innovations developed by Caltech and NASA's Jet Propulsion Laboratory.

Andrew Adonis was Secretary of State for Transport, Chairman of the UK National Infrastructure Commission and head of the Number 10 Policy Unit think tank.

Ben Alun-Jones is a co-founder and Creative Director of Unmade, the leading customisation platform for the fashion industry. His aim is to directly connect consumer choices to automated manufacturing.

Alejandro Aravena is an architect best known for his innovative social housing schemes. He is the Executive Director of the firm Elemental S.A. and in 2016 won the Pritzker Architecture Prize and was director and curator of the Venice Architecture Biennale.

John Armitt is a civil engineer. Among other positions, he has been President of the Institution of Civil Engineers and Chairman of the Olympic Delivery Authority, which was responsible for the London Olympic Games in 2012. He is also Chairman of National Express Group and City & Guilds.

Nicholas Ashton is a Curator of the Palaeolithic and Mesolithic Collections at the British Museum. He specialises in the earliest human occupation of Northern Europe and has directed and published research on several major excavation projects.

John Bell is an immunologist and a leader in the field of genomic science. He was President of the Academy of Medical Sciences and is currently Regius Professor of Medicine at the University of Oxford.

Robin Bendrey is a lecturer in Archaeology at the University of Edinburgh. His research is focused on the origins, development and impact of animal husbandry and its use in prehistoric Eurasia.

Tim Berners-Lee invented the World Wide Web in 1989. He is a winner of the Queen Elizabeth Prize for Engineering and the Turing Award, the most prestigious prizes in Engineering and Computer Science.

Tilly Blyth is Head of Collections and Principal Curator at the Science Museum in London. She is also a trustee of the Raspberry Pi Foundation, a UK-based educational charity that works to put the power of computing and digital making into the hands of people all over the world.

Ed Boyden is a neuroscientist and a bioengineer at the Massachusetts Institute of Technology. He is a faculty member in the MIT Media Lab, an associate member of the McGovern Institute for Brain Research and has invented several technologies that advance brain science.

Nigel Brandon is the Dean of Engineering at Imperial College London. He was the founding director of the Energy Futures Lab there and also founded the AIM-listed fuel cell company, Ceres Power. His research focus is on electrochemical power sources for fuel cell and energy storage applications.

Cynthia Breazeal is an Associate Professor of Media Arts and Sciences at the Massachusetts Institute of Technology, where she founded and directs the Personal Robots Group at the Media Lab. She is also the founder and Chief Scientist of Jibo Inc., which develops social robots and robotic assistants.

Clive Brown is Chief Technology Officer at Oxford Nanopore Technologies. He leads the specification and design of the company's nanopore-based sensing platform, which provides new ways to study DNA, protein and other biomolecules.

Sue Brunning is Curator of the European Early Medieval and Sutton Hoo Collections at the British Museum. She specialises in early Anglo-Saxon material culture.

Nick Butler is Visiting Professor and Chair of the King's Policy Institute at King's College London. He spent twenty-nine years working at BP, including five years as Group Vice President for Policy and Strategy Development. He has also been Senior Policy Adviser to the UK Prime Minister, Chairman of the Centre for European Reform and Treasurer of the Fabian Society.

Simon Calvert is Head of Product Definition and Innovation and Chief Technology Officer at Siemens Magnet Technology.

Tristram Carfrae is Deputy Chairman of Arup and a leading structural designer, who is responsible for many award-winning buildings, including the Beijing National Aquatics Center.

Roger Carr is Chairman of BAE Systems, Vice-Chairman of the BBC Trust and a senior advisor to KKR. He is also a Visiting Fellow of Saïd Business School at the University of Oxford.

Vint Cerf is recognised as one of 'the fathers of the Internet'. A winner of the Queen Elizabeth Prize for Engineering, he is also currently a Vice President and Chief Internet Evangelist at Google.

Suranga Chandratillake is the founder and President of blinkx, an Internet video search engine. He is also a General Partner at the London-based Venture Capital firm Balderton Capital.

George Church leads Synthetic Biology at the Wyss Institute at Harvard University and is a Professor at the Massachusetts Institute of Technology. In his career he has pioneered new techniques in genome sequencing, editing, writing and recoding.

Rory Collins is a Professor of Medicine and Epidemiology at the University of Oxford, and was appointed Principal Investigator and Chief Executive of UK Biobank in 2005.

Lucy Collinson is the lead researcher for the Electron Microscopy Science Technology Platform at the Francis Crick Institute in London.

Oliver Cooke is a Curator of Horology at the British Museum, where he is responsible for the conservation of clock movements, to ensure that the collection maintains a stable condition.

Steven Cowley is a theoretical physicist who specialises in nuclear fusion and astrophysical plasmas. He is Director of the US Department of Energy Princeton Plasma Physics Laboratory and was previously President of Corpus Christi College at the University of Oxford and Chief Executive Officer of the United Kingdom Atomic Energy Authority.

Diane Coyle is an economist and a former adviser to the UK Treasury. She was also Vice-Chairman of the BBC Trust, a member of the UK Competition Commission and is currently a professor at the University of Cambridge.

Missy Cummings is an Associate Professor at Duke University and Director of Duke's Humans and Autonomy Laboratory. A naval officer and military pilot 1988–99, she was also one of the US Navy's first female fighter pilots.

Ara Darzi is one of the world's leading surgeons. He holds the Paul Hamlyn Chair of Surgery at Imperial College London and specialises in the field of minimally invasive and robot-assisted surgery, having pioneered many new techniques and technologies.

Sally Davies has been Chief Medical Officer for England and Chief Medical Advisor to the UK government since 2011. She was also a member of the Executive Board of the World Health Organization, 2014–16. Her clinical specialism was haematology.

Robert Davis is an archaeologist in the British Museum's Department of Britain, Europe and Prehistory.

Ryan Ding is Executive Director and President of Products & Solutions at Huawei. He was previously Chief Executive Officer of the Carrier Network Business Group.

Ann Dowling is President of the Royal Academy of Engineering. She is also a Professor of Mechanical Engineering at the University of Cambridge, where she was previously Head of the Department of Engineering.

Eric Drexler is a researcher and author in the field of nanotechnology. He is Senior Research Fellow at the Future of Humanity Institute, and Oxford Martin Senior Fellow at the Oxford Martin School, University of Oxford.

Simon Duckett is a Professor of Chemistry at the University of York, where he is also Director of the Centre for Hyperpolarisation in Magnetic Resonance.

Mike Dunne is Director of the Linac Coherent Light Source at Stanford University. He is also an Associate Laboratory Director of the SLAC National Accelerator Laboratory and a Professor of Photon Science at Stanford University.

Hugh Durrant-Whyte is Chief Scientific Adviser at the UK Ministry of Defence. He is also known for his pioneering work on probabilistic methods for robotics.

Tony Fadell was previously Senior Vice President of the iPod division at Apple and is known as 'one of the fathers of the iPod'. He is also founder of Nest Labs and led the Google Glass division at Alphabet Inc. 2015–16.

Jeremy Farrar is Director of the Wellcome Trust and before that was Professor of Tropical Medicine at the University of Oxford. He has also served as chair on multiple advisory boards for governments and global organisations, including the World Health Organization.

Greg Flanagan is the Head of Manufacturing at BAE Systems.

Eric Fossum is the inventor of the MOS active pixel image sensor and the CMOS image sensor. He is currently a Professor at the Thayer School of Engineering at Dartmouth College and is also a winner of the Queen Elizabeth Prize for Engineering.

Norman Foster is a world-renowned architect. He is best known for his work on 'The Gherkin' in London, Apple Park in Cupertino and the headquarters of HSBC in Hong Kong. He holds the Pritzker Architecture Prize and the Prince of Asturias Award.

Paul Freemont is Head of the Section of Structural Biology within the Department of Medicine at Imperial College London. He is also co-founder and Co-Director of the EPSRC Centre for Synthetic Biology and Innovation and the National UK Innovation and Knowledge Centre for Synthetic Biology.

Richard Friend holds the Cavendish Professorship of Physics at the University of Cambridge. His research covers the physics, materials science and engineering of organic semiconductors. He is also Director of the Winton Programme for the Physics of Sustainability and of the Maxwell Centre. He co-founded Eight19 to commercialise solar photovoltaic technologies invented in his lab.

Steve Furber is the ICL Professor of Computer Engineering at the School of Computer Science at the University of Manchester. He is best known for his work at Acorn Computers, where he was one of the designers of the BBC Micro and the ARM 32-bit RISC microprocessor.

James Garvin is Chief Scientist at the NASA Goddard Space Flight Center. His academic expertise is in earth and planetary sciences, and he provides strategic advice and analysis on scientific priorities to NASA leadership.

Zoubin Ghahramani is Professor of Information Engineering at the University of Cambridge. He is an expert on machine learning and artificial intelligence and holds joint appointments at Carnegie Mellon University, University College London and the Alan Turing Institute. He is also Chief Scientist at Uber and Deputy Director of the Leverhulme Centre for the Future of Intelligence.

Sam Goldman is founder of d.light design, a company established to develop and deliver affordable solar-powered solutions for people without access to reliable electricity networks.

Antony Griffiths held the Slade Professorship of Fine Art at the University of Oxford and was also Keeper of the Department of Prints and Drawings at the British Museum between 1991 and 2011.

Caroline Hargrove is Technical Director of McLaren Applied Technologies. She is best known for her simulation of human–machine interaction, initially in motor racing and subsequently in other sporting and medical applications.

Claire Harris is an archaeologist in the British Museum's Department of Britain, Europe and Prehistory.

Peter Head is founder and Chief Executive Officer of the Ecological Sequestration Trust and founder and Chair of the Resilience Brokers programme. He was previously a Director of Arup and was also appointed as Independent Commissioner on the London Sustainable Development Commission in 2002.

Thomas Heatherwick is a designer and the founder of London-based design practice Heatherwick Studio. The studio's prolific output defies conventional classification and is characterised by ingenuity, inventiveness and originality.

John Hennessy is a computer scientist, academic and businessman. He served as a President of Stanford University and is one of the founders of MIPS Computer Systems Inc. and Atheros. He was awarded the Turing Award in 2018 for pioneering a new approach for designing faster, lower power and reduced complexity microprocessors. He is currently the Executive Chairman of Alphabet Inc.

Peter Hennessy is Attlee Professor of Contemporary British History at Queen Mary University of London. He has written more than a dozen books on modern British history, and has won prizes including the Orwell Prize and the Duff Cooper Prize. He was made a life peer in 2010.

Dave Holmes is Director of Manufacturing Operations for the Military Air and Information Business Unit at BAE Systems.

Andrew Hopper is the Professor of Computer Technology and Head of the University of Cambridge Computer Laboratory. He is best known for co-founding the technology firm Acorn, the company that built the BBC Microcomputer.

Colin Humphreys is the current Director of Research and a Fellow of Selwyn College at the University of Cambridge. His research focuses on gallium nitride semiconductors, advanced electron microscopy and developing high-temperature aerospace materials. He was the founder and Director of the companies CamGaN and Intellec.

Mark James is Chief Technology Officer and Executive Director of the cognitive computing company Beyond Limits. He was a Principal Investigator at NASA's Jet Propulsion Laboratory, where he developed his expertise in artificial intelligence and machine learning.

Calestous Juma was, until his death in 2017, Professor of the Practice of International Development at Harvard University. He was an internationally recognised authority in the application of science and technology to sustainable development worldwide.

Dave Kashen is co-founder of Fearless Ventures, a venture capital firm committed to helping entrepreneurs become better leaders and build businesses for the benefit of all stakeholders.

Nicholas Kelez is Chief Engineer and Mechanical Engineering Division Director at SLAC National Accelerator Laboratory.

Brian Krzanich was the Chief Executive Officer of Intel until June 2018, having joined the company in 1982. He is also a member of the board of directors of the Semiconductor Industry Association and the US Drone Advisory Committee.

Robert Langer is a chemical engineer, scientist, entrepreneur and inventor. He is a winner of the Queen Elizabeth Prize for Engineering and is currently the David H. Koch Institute Professor at the Massachusetts Institute of Technology. He has written more than 1,400 articles and has over 1,260 issued and pending patents worldwide.

Benedict Leigh is Curator of the Ancient Near East Collection at the British Museum.

Amanda Levete is a RIBA Stirling Prize-winning architect and founder and Principal of AL_A, an international award-winning design and architecture studio.

Fei-Fei Li is Professor of Computer Science at Stanford University and the Director of the Stanford Artificial Intelligence Lab and the Stanford Vision Lab. She was Chief Scientist of Artificial Intelligence and Machine Learning at Google Cloud 2017–18.

Michael Mayberry is Chief Technology Officer for Intel Corporation and Senior Vice President and Managing Director of Intel Labs. He is responsible for Intel's global research efforts in computing and communications.

Jianjun Mei is Director of the Needham Research Institute and a Fellow of Churchill College at the University of Cambridge. He specialises in the origins and role of metallurgy in Early China, and cultural interactions between China and the West.

Mike Menzel is the NASA Mission Systems Engineer for the James Webb Space Telescope at the Goddard Space Flight Center. Before joining NASA in 2004, he worked with Northrop Grumman and Lockheed Martin.

Seth Miller is co-founder of Fearless Ventures, a venture capital firm committed to helping entrepreneurs become better leaders and build businesses for the benefit of all stakeholders.

Onnig Minasian is a computer systems engineer who specialises in design, integration, performance, security and availability of wide area networks, enterprise private networks and very large distributed databases.

Justin Molloy is the Lead Researcher of the Single Molecule Enzymology Laboratory at the Francis Crick Institute.

Ernie Moniz is the Cecil and Ida Green Professor of Physics and Engineering Systems at the Massachusetts Institute of Technology. He was US Secretary of Energy 2013–17. He is currently Co-Chairman and Chief Executive Officer of the Nuclear Threat Initiative, which works to prevent catastrophic attacks with weapons of mass destruction and disruption, and is also CEO of the non-profit Energy Futures Initiative.

Paul Newman is the BP Professor of Information Engineering and head of the Oxford Robotics Institute at the University of Oxford and an EPSRC Leadership Fellow. He is co-founder of the autonomous vehicle software firm Oxbotica.

Paul Nurse is an English geneticist, a former President of the Royal Society and Chief Executive and Director of the Francis Crick Institute. He was jointly awarded the 2001 Nobel Prize in Physiology or Medicine.

Onora O'Neill is a philosopher and member of the House of Lords. She is an Emeritus Professor of Philosophy at the University of Cambridge, a former President of the British Academy and chaired the Nuffield Foundation.

Toby Ord is a moral philosopher at Oxford University, where he is a Research Fellow at the Future of Humanity Institute. In 2009 he launched Giving What We Can, an international society whose members each pledge to donate at least 10 per cent of their income to anti-poverty charities.

Brad Parkinson was the Chief Architect for the Global Positioning System (GPS) and led the design and implementation of the project from its inception in 1973. He was appointed a Professor at Stanford University in 1984.

Ben Reed is Deputy Division Director in the Satellite Servicing Projects Division at NASA Goddard Space Flight Center. He is also a Senior Policy Adviser at the National Space Council.

Ralph Robins was Chief Executive Officer and later Chairman of Rolls-Royce. He joined the company as a graduate apprentice in 1955 and over his career worked on numerous innovations in jet engine engineering.

Richard Rogers is a world-renowned architect. He is best known for his work on the Pompidou Centre in Paris, the Lloyd's building and the Millennium Dome in London and the European Court of Human Rights building in Strasbourg. He is a winner of the RIBA Gold Medal, the Thomas Jefferson Medal, the RIBA Stirling Prize and the Pritzker Prize.

David Rooney was Keeper of Technologies and Engineering at the Science Museum in London until 2018. He is also an Honorary Research Associate at Royal Holloway, University of London.

Antoine Rostand is the founder and President of Kayrros, a Paris-based start-up which uses machine learning approaches to provide analytics for energy markets. Previously, he was President and founder of Schlumberger Business Consulting.

Daniela Rus is the Andrew and Erna Viterbi Professor of Electrical Engineering and Computer Science at the Massachusetts Institute of Technology, where she is also Director of the Computer Science and Artificial Intelligence Laboratory. Her research interests are in robotics, artificial intelligence and data science.

Ben Russell is Curator of Mechanical Engineering at the Science Museum in London. He has curated several major exhibitions at the museum, including 'Robots' (2017), 'Cosmonauts' (2015) and 'James Watt's Workshop' (2011).

Stuart Russell is Professor of Computer Science and the Smith-Zadeh Professor in Engineering at the University of California, Berkeley. He is known for his contributions to artificial intelligence. His current concerns include the threat of autonomous weapons and the long-term future of artificial intelligence and its relation to humanity.

Justin Sanchez is Director of the Biological Technologies Office at DARPA, where he oversees ten programmes in areas of science and technology development including neuro-technology, gene editing, synthetic biology and outpacing infectious diseases. He was an Associate Professor of Biomedical Engineering and Neuroscience at the University of Miami.

Anders Sandberg is a James Martin Research Fellow at the Future of Humanity Institute at the University of Oxford. He is known for his work on societal and ethical issues surrounding human enhancement and new technology.

John Savill is Regius Professor of Medical Science at the University of Edinburgh. Until 2018 he was Chief Executive of the Medical Research Council in the UK.

Nigel Shadbolt is Principal of Jesus College, Oxford, and Professorial Research Fellow in the Department of Computer Science at the University of Oxford. He is also Chairman of the Open Data Institute, which he founded with Tim Berners-Lee.

Amnon Shashua is co-founder, Chief Executive Officer and Chief Technology Officer of Mobileye, which develops computer vision-based software systems for advanced driver assistance and autonomous vehicles. He is also Vice President at Intel and a Computer Science Professor at the Hebrew University in Jerusalem.

Hadar Shemtov is Director of Engineering at Google, where he specialises in Search and Relevance. He has fifteen patents related to search algorithms and other technologies.

Pontus Skoglund is a population geneticist and a group leader at the Francis Crick Institute, where he studies ancient DNA to better understand early human history.

Mike Spence is an economist and joint recipient of the 2001 Nobel Prize in Economic Sciences. He was formerly the Dean of the Graduate School of Business at Stanford University.

Phil Stride is Strategic Projects Director of Tideway, which is constructing the Thames Tideway Tunnel, a major new sewer for London.

Martin Sweeting is the founder and Executive Chairman of Surrey Satellite Technology, a company that builds and operates small satellites. He is also Distinguished Professor of Space Engineering at the University of Surrey.

Rahim Tafazolli is Professor of Mobile and Satellite Communications at the University of Surrey. He is also Director of the Institute of Communication Systems and Director of the 5G Innovation Centre.

Kevin Tebbit was Permanent Under-Secretary of State at the UK Ministry of Defence. He has also served as Director of GCHQ, and is a Visiting Professor at Queen Mary University of London.

Nobukazu Teranishi is known for inventing the pinned photodiode, which is used in most digital cameras today. He is a Professor at the University of Hyogo and at the University of Shizuoka, and is a winner of the Queen Elizabeth Prize for Engineering.

Justin Vale is a Consultant Urological Surgeon at the Imperial College Healthcare NHS Trust in London, where he specialises in laparoscopic and robotic surgery.

Craig Venter is a biotechnologist, biochemist, geneticist and businessman. He is best known for his involvement with the early human genome sequencing efforts and also assembled the first team to create a living cell with entirely synthetic chromosomes.

Paul Verhoef is Director of the Galileo Programme and Navigation-related Activities at the European Space Agency.

Mike Walker is a Principal Researcher at Microsoft, where he works on security AI. He is also a Project Manager at DARPA, where he led DARPA's Cyber Grand Challenge, a two-year contest to construct and compete the first prototypes of reasoning cyber-defence AI.

Mark Walport is Chief Executive of UK Research and Innovation (UKRI), which is responsible for the public funding of research and innovation. He was previously UK Government Chief Scientific Adviser and, before that, Director of the Wellcome Trust.

Paul Westbury is Group Technical Director at engineering practice Laing O'Rourke. He previously served as Chief Executive Officer of engineering firm Buro Happold.

Tim Westergren is a co-founder of Pandora, a music streaming and automated music recommendation service.

Nigel Whitehead is Chief Technology Officer at BAE Systems.

Ed Williams is CEO and President of the UK and Ireland arm of Edelman, a global communications marketing firm. He previously worked in senior communications positions at the BBC and Reuters.

Dan Yakir is a Professor of Earth and Planetary Sciences at the Weizmann Institute of Science in Israel. His research focuses on ecosystem-wide energy budgets and the interactions between the biosphere and atmosphere during climate change.

Kenichi Yoshida is Chief Business Officer, Executive Vice President and Vice President of Business Development at SoftBank Robotics. He was previously a co-founder and Chief Operating Officer of Realcom Inc., an enterprise software start-up.

Dieter Zetsche is Chairman of the Board of Management of Daimler AG and Head of Mercedes-Benz Cars.

Tong Zhang is a machine-learning researcher, and the Executive Director of Tencent AI Lab. Previously, he was a professor at Rutgers University, and worked at IBM, Yahoo and Baidu.

LIST OF FIGURES

CHAPTER 3: THINK

temperature. *Original artwork by Edelman, based on data generously provided by Climate Central.*

NOTES

1 PROGRESS

1 'Global Health Check' by the Bill & Melinda Gates Foundation and Mosaic. Available online at https://mosaicscience.com/story/global-health-check

2 Max Roser, 'Child Mortality', 2018, which summarises data from the World Bank and Gapminder. Available online at https://ourworldindata.org/child-mortality

3 Max Roser, 'Life Expectancy', 2018, which summarises historical data from Clio Infra and the United Nations Population Division. Available online at https://ourworldindata.org/life-expectancy

4 J. B. DeLong, 'Estimating World GDP, One million BC–Present', draft paper, Berkeley, California, 1998. And CIA World Factbook.

5 Max Roser and Esteban Ortiz-Ospina, 'Global Extreme Poverty', 2018, which summarises data from the World Bank. Available online at https://ourworldindata.org/extreme-poverty. Also François Bourguignon and Christian Morrisson, 'Inequality Among World Citizens: 1820–1992', in American Economic Review, 92(4), pp. 727–48.

6 Max Roser and Esteban Ortiz-Ospina, 'Literacy', 2018, which summarises data from the Organisation for Economic Co-operation and Development and UNESCO. Available online at https://ourworldindata.org/literacy

7 Loss aversion and the availability bias are described in detail in Daniel Kahneman's Thinking, Fast and Slow, Macmillan, New York, 2011.

8 A recent opinion poll, conducted by YouGov in seventeen different countries, revealed widespread pessimism about the state of the world. Of the countries sampled, China was the only one where a majority of people perceived the state of the world to be improving. See https://yougov.co.uk/news/2016/01/05/chinese-people-are-most-optimistic-world

9 This statement was written by Facebook Vice President Andrew Bosworth in 2016. N. Bowles and S. Frenkel, 'Facebook Employees

in an Uproar Over Executive's Leaked Memo', *New York Times*, 30 March 2018.

10   Ida M. Tarbell, *The History of the Standard Oil Company*, Cosimo, Inc., 2010.

2 MAKE

1   S. Harmand, J. E. Lewis, C. S. Feibel, C. J. Lepre, S. Prat, A. Lenoble, X. Boës, R. L. Quinn, M. Brenet, A. Arroyo and N. Taylor, '3.3-million-year-old stone tools from Lomekwi 3, West Turkana, Kenya' *Nature*, 521(7552), 2015, pp. 310–15.

2   Despite the relatively limited repertoire of tools available, it is worth noting the great influence that early humans had on the planet. They went from marginal primates living in isolated pockets of Africa to colonising every continent and every available habitat. They evolved from scraping together a meagre diet of carrion and foraged plants to become highly skilled hunter-gatherers. Tens of millennia before the invention of organised agriculture, people were actively re-shaping whole ecosystems. At first, people must have lived in constant fear of large carnivores, but over millennia they came to occupy the top tier of the food chain. The fossil record shows that everywhere people went, waves of mass extinction spread through the animal populations they encountered. No other life form has ever reinvented its ecology so rapidly or so wholly. The hand axe and a crude mastery of fire opened the world to our deep ancestors. Read more about this in Yuval Noah Harari, *Sapiens*, 2015, Vaclav Smil, *Energy and Civilisation*, 2017 and Jared Diamond, *Guns, Germs and Steel*, 1998.

3   K. Alder, 'Innovation and Amnesia: Engineering Rationality and the Fate of Interchangeable Parts Manufacturing in France', *Technology and Culture*, 38(2), 1997, pp. 273–311.

4   Adam Smith, *An Inquiry into the Nature and Causes of the Wealth of Nations*, 1776.

5   I refer to the measures of economic output and growth that are collated and interpreted by Professor Max Roser of Oxford University. See Max Roser, 'Economic Growth', 2018. Available online at https://ourworldindata.org/economic-growth

6   *Ericsson Mobility Report*, Mobile World Congress, Stockholm, 2016.

7 Joseph W. Roe, *English and American Tool Builders,* Yale University Press, 1916. And *The Report of the British Commissioners to the New York Industrial Exhibition,* London, 1854.

8 Scientific management emerged towards the end of the nineteenth century and was the first and most influential attempt to analyse systematically and then optimise industrial processes and workflows. The approach was advanced first by Frederick W. Taylor, a foreman at the Midvale Steel Works in Pennsylvania, USA during the 1880s. Taylor, emphasising the need for empirical analysis, timed and optimised the different steps executed by workers in factory processes to improve economic efficiency. A little later, Frank and Lillian Gilbreth introduced 'time and motion studies', which were designed to optimise the physical movements made by workers, with the aim of improving both production efficiency and worker welfare.

9 G. N. Georgano, *Cars: Early and Vintage, 1886–1930,* Grange-Universal, London, 1985. And http://corporate.ford.com/innovation/100-years-moving-assembly-line.html

10 The Robotic Industries Association, *A Tribute to Joseph Engelberger,* 2018. Available online at https://www.robotics.org/joseph-engelberger/unimate.cfm

11 Genetically engineered human insulin was first produced in 1978, after scientists inserted the human insulin gene into *E. coli* bacteria. It was approved for medical use in 1982. Today almost all the insulin used by diabetics is produced in a similar way. It is safer, identical to naturally occurring insulin and produced very cheaply and at scale. Before this was possible, diabetics relied on insulin that was derived from animals, usually pigs and cows – it was expensive and carried a small risk of contamination. Another example of an important class of products that depends on the large-scale industrial production of engineered cells is biotherapeutic monoclonal antibodies, which are used to treat a range of diseases including some cancers and autoimmune conditions such as rheumatoid arthritis, Crohn's disease and ulcerative colitis.

12 M. Bon, 'A Discourse upon the Usefulness of the Silk of Spiders', *Philosophical Transactions (1683–1775),* 27, pp. 2–16.

13 J. Forster, M. Bryan and A. Stone, 'Doing Whatever a Spider Can', *Physics Special Topics,* 11(1), 2012.

14   A. Z. Nielsen, S. B. Mellor, K. Vavitsas, A. J. Wlodarczyk, T. Gnanasekaran, M. Perestrello Ramos H de Jesus, B. C. King, K. Bakowski and P. E. Jensen, 'Extending the biosynthetic repertoires of cyanobacteria and chloroplasts', *The Plant Journal*, 87(1), pp. 87–102. Also I. Ajjawi, J. Verruto, M. Aqui, L. B. Soriaga, J. Coppersmith, K. Kwok, L. Peach, E. Orchard, R. Kalb, W. Xu and T. J. Carlson, 'Lipid production in Nannochloropsis gaditana is doubled by decreasing expression of a single transcriptional regulator', *Nature Biotechnology*, 35(7), 2017, p. 647.

15   G. O. Chen, X. R. Jiang, and Y. Guo, 'Synthetic biology of microbes synthesizing polyhydroxyalkanoates (PHA)', *Synthetic and Systems Biotechnology*, 1(4), 2016, pp. 236–42.

16   The US biotechnology sector is one of the fastest-growing parts of the economy. In total, biotechnology is already estimated to bring in annual revenue of at least $324 billion. This is currently just 2 per cent of the country's GDP, but that share is expanding by 10 per cent each year. Figures from R. Carlson, 'Estimating the biotech sector's contribution to the US economy', *Nature Biotechnology*, 34(3), 2016, p. 247.

17   The synthetic biology community has learned from the earlier communication errors that stoked the public fear of genetic engineering. Most major research programmes now involve social scientists and ethicists to forecast and respond to the social impacts and public concerns about the new technology. Moreover, synthetic biologists are devising ways to hard-code safety and control features into re-engineered cells. Harvard's Professor George Church, one of the pioneers in the field of synthetic biology, describes himself as a 'professional worrier', but one who converts his anxiety into pre-emptive action. In 2015, for example, he created strains of engineered bacteria that are totally reliant for growth upon a chemical that is not found in nature. This dependency amounts to a 'kill switch': there is no possibility of these bacteria surviving outside the lab or factory where they are used. See D. J. Mandell, M. J. Lajoie, M. T. Mee, R. Takeuchi, G. Kuznetsov, J. E. Norville, C. J. Gregg, B. J. Stoddard and G. M. Church, 'Biocontainment of genetically modified organisms by synthetic protein design', *Nature*, 518(7537), 2015. pp. 55–60.

18   T. Hancock, 'Adidas boss says large-scale reshoring is "an illusion"', *Financial Times*, 23 April 2017.

19 For example, see R. D. Sochol, E. Sweet, C. C. Glick, S. Y. Wu, C. Yang, M. Restaino and L. Lin, '3D printed microfluidics and microelectronics', *Microelectronic Engineering*, 2017.

20 ING Group, '3D Printing: A Threat to Global Trade', 2017.

21 If large-scale 'reshoring' of manufacturing does occur, it could create new challenges for global development. Until very recently, industrial manufacturing gravitated to places where labour is cheap, usually in developing markets. If automated manufacturing, based on techniques like 3D printing, no longer relies upon low-cost labour, countries that are at an early stage of development would likely suffer most. If low-cost manufacturing is no longer a key asset, 'What is the growth model for early-stage developing countries?' Spence asks. This is a question that needs to be deliberated upon, though it is far too early to be alarmed.

22 T. Hancock, 'Adidas boss says large-scale reshoring is "an illusion"', *Financial Times*, 23 April 2017.

23 Drexler helped to launch the field of nanotechnology in the 1980s. He lodged his ideas in the public awareness with his popular book, *The Engines of Creation: The Coming Era of Nanotechnology* (1986). The US government alone has spent more than $24 billion on nanotechnology research since 2001, although this has been focused on materials science and the engineering of nanoscale devices, rather than on developing the kind of molecular manufacturing proposed by Drexler. In 2015 the US Department of Energy hosted a major workshop on the topic of atomically precise manufacturing. 'The outcome was great. It ended up with proposals that look productive', Drexler told me. 'But then Trump came in, and I don't know what is going on', he added.

24 R. F. Service, 'How to build a better battery through nanotechnology', *Science*, 27 May 2016.

25 This research paper shows the proof of principle for a transistor just one nanometre long. S. B. Desai, S. R. Madhvapathy, A. B. Sachid, J. P. Llinas, Q. Wang, G. H. Ahn, G. Pitner, M. J. Kim, J. Bokor, C. Hu and H. S. P. Wong, 'MoS_2 transistors with 1-nanometer gate lengths', *Science*, 354(6308), 2016, pp. 99–102.

26 David Ricardo, *On the Principles of Political Economy and Taxation*, John Murray, London, 1817.

27 James Bessen, *Learning by Doing: The Real Connection Between Innovation, Wages, and Wealth*, Yale University Press, New Haven, 2015.

28 'Gartner Worldwide, IT Spending, Worldwide, 1Q18 Update', Gartner, Inc. Stamford, 2018. And 'Evans Data Corporation, Global Developer Population and Demographic Study 2017 Vol. 2', Evans Data Corporation, Santa Cruz, 2017.

29 James E. Bessen, 'How Computer Automation Affects Occupations: Technology, Jobs, and Skills', Boston University School of Law and Economics Research Paper, 3 October 2016.

30 Ibid.

31 Steven Pinker, *Enlightenment Now: The Case for Reason, Science, Humanism, and Progress*, Penguin, London, 2018, p. 475

32 M. Huberman and C. Minns, 'The times they are not changin': Days and hours of work in Old and New Worlds, 1870–2000', *Explorations in Economic History*, 44(4), 2007, pp. 538–67.

33 J. M. Keynes, 'Economic possibilities for our grandchildren' in *Essays in Persuasion*, 1933, pp. 358–73.

34 Robert J. Gordon, *The Rise and Fall of American Growth: The U.S. Standard of Living Since the Civil War*, Princeton University Press, Princeton, 2016.

35 Robert M. Solow, 'We'd better watch out', *New York Times Book Review*, 12 July 1987.

36 You can read more about the introduction of the electric dynamo in P. A. David, 'The Dynamo and the Computer: An Historical Perspective on the Modern Productivity Paradox', *The American Economic Review*, 80(2), 1990, pp. 355–61.

37 Replacing entire systems is usually prohibitively expensive and comes with a risk of outright failure. Instead, both software and hardware tend to be added to existing systems incrementally, causing the overall level of complexity to rise steeply. These legacy systems can be brittle and exhibit unpredictable behaviour when they are altered. 'You can't just press the rest button on these systems,' Hopper explains. 'This,' he concludes, 'is a good reason why the manufacturing people say they won't take a risk.'

38 L. Mishel and J. Bivens, 'The zombie robot argument lurches on', *Economic Policy Institute*, 24 May 2017.

39 J. Manyika, *Manufacturing the future: The next era of global growth and innovation*, McKinsey Global Institute, 2012.

40 This comment and some of the other quotes from Sweeting come from a speech that he gave at the Royal Aeronautical Society in 2017. Transcript available online at https://www.sstl.co.uk/Blog/May-2017/Sir-Martin-s-Royal-Aeronautical-Society-Annual-Din

41 'World's first "phonesat", STRaND-1, successfully launched into orbit', Surrey Satellite Technology Ltd, 25 February 2013. Available online at http://www.sstl.co.uk/Press/Worlds-first-phonesat-STRaND-1-successfully-launched

42 V. Smil, *Making the Modern World: Materials and Dematerialization*, John Wiley & Sons, London, 2016.

43 See Reep Technologies Ltd. http://reepcorp.com

44 S. Yoshida, K. Hiraga, T. Takehana, I. Taniguchi, H. Yamaji, Y. Maeda, K. Toyohara, K. Miyamoto, Y. Kimura and K. Oda. 'A bacterium that degrades and assimilates poly(ethylene terephthalate)', *Science*, 351(6278), 11 March 2016, pp. 1196–9.

45 M. Andreessen, 'Why software is eating the world', *Wall Street Journal*, 20 August 2011.

3 THINK

1 Most of the details about the Nazis' use of tabulating machines come from Edwin Black, *IBM and the Holocaust: The Strategic Alliance between Nazi Germany and America's Most Powerful Corporation*, Dialog Press, New York, 2012.

2 Edwin Black provides evidence that IBM's senior management, including their chairman Thomas Watson, knew about this trade and continued to provide technical support to Nazi customers after they had learned how the machines were being used.

3 Georges Ifrah, *The Universal History of Computing: From the Abacus to the Quantum Computer*, John Wiley & Sons, New York, 2001, p. 11.

4 In 1900, Greek divers investigating the remains of a ship that sank in 65 BC discovered the Antikythera mechanism, sometimes described as the 'world's first computer'. Study of the corroded bronze device revealed a complex clockwork mechanism of at least thirty-seven toothed gears. The device would have displayed the relative positions of the sun, the moon and all of the five planets visible to the naked eye (Mars, Venus, Mercury, Jupiter and Saturn) as well as the phase of the moon. It had the ability to calculate the positions of these celestial bodies and to predict

eclipses and other astronomical and astrological phenomena years into the future. Although technically a calculator rather than a computer, no known device of similar age remotely matches the technological sophistication of the Antikythera mechanism.

5 Babbage in November 1839, recalling events in 1821. Quoted in Harry Wilmot Buxton and Anthony Hyman, *Memoir of the Life and Labours of the Late Charles Babbage*, MIT Press, Cambridge, 1988.

6 Calculating each figure in a nineteenth-century mathematical table usually required several sequential steps, and the results then had to be transcribed for printing. Finally, the printers had to set the type manually. At every stage of the process there was a risk of mistakes. Without repeating the complex mathematics oneself, or waiting to see if your ship laden with precious cargo would run aground, there was no way of knowing exactly how trustworthy any particular number in a table was. Another friend and contemporary of Babbage and Herschel, Dionysius Lardner, surveyed a set of forty randomly selected mathematical tables available at the time – he found 3,700 errors within and predicted that were many more besides. See http://www.rutherfordjournal.org/article030106.html

7 Probably the most lastingly influential written work of Classical Greek astronomy is Ptolemy's *Almagest*, written in the second century AD. In this vast, multi-volume work, Ptolemy summarised and extended the work of previous generations of Greek, Byzantine and Mesopotamian astronomers and astrologers. Although added to by later scholars, *Almagest* remained a core reference point for astronomers throughout India, the Middle East and Europe until Copernicus finally overturned the geocentric view of the cosmos in the sixteenth century.

8 After more than a decade of hard work on the Difference Engine, Babbage's lead engineer, Joseph Clement, walked away – it seems that he was no longer prepared to work with the bristly and argumentative Babbage. At the moment that the engineer abandoned the project, 12,000 of the estimated 25,000 parts of Babbage's contraption had been painstakingly cast and machined and Babbage had spent £17,000 of government money, the cost of two contemporary battleships. Soon afterwards, the bulk of his state funding was withdrawn.

9 The team built the machine to Babbage's precise specifications, using only materials and production tolerances that were available in the

nineteenth century; this dispels the idea that Babbage's projects fell apart because his vision exceeded the engineering capabilities available to him at the time. More likely, his irritable personality and deplorable project-management skills derailed his ambitions.

10 Charles Babbage, *Passages from the Life of a Philosopher*, Longman, London, 1864.

11 Babbage co-opted the idea of using punched paper cards to program his numerical engines from the late eighteenth century 'Jacquard' silk loom. According to their paper-born instructions, Jacquard looms executed complex weaving patterns with a precision that couldn't be matched by unassisted human weavers.

12 One key feature of the Analytical Engine, repeated in computers today, was the fact that it had separate storage and processing mechanisms. Babbage called them the 'mill' and the 'store'. Just like in the workings of a modern microprocessor, information is stored in memory and brought to the central processing unit only as and when it is needed – these are called 'fetch-execute' cycles. The Analytical Engine would have been capable of iterative looping, conditional branching and several other key algorithmic operations that will be familiar to anyone who has dipped into the world of computer programming.

13 The letters shared between Lovelace and Babbage are archived at the British Library. Available online at http://www.bl.uk/collection-items/letter-from-ada-lovelace-to-charles-babbage

14 In 1842 Lovelace translated a paper written by the Italian engineer Luigi Menabrea, which described an early version of the Analytical Engine. As she made her translation, she added extensive notes of her own. Within these notes is the first computer algorithm that she, or anyone else, had composed. It was designed to calculate Bernoulli numbers, the still-somewhat-mysterious series of rational numbers that have an important place in several branches of mathematics. Lovelace clearly describes the specific details of seventy-five 'variable cards' that would be needed to program the analytical engine, and points out how they could be reused cyclically (what we'd now call a programming loop), to compute the whole number series. You can read Ada Lovelace's translation and notes online at https://www.fourmilab.ch/babbage/sketch.html

15 Ada Lovelace expresses her critical new concept of abstraction in another annotation to her 1842 translation: '[The Analytical Engine] might act upon other things besides numbers, were objects found whose mutual fundamental relations could be expressed by those of the abstract science of operations, and which should be also susceptible of adaptations to the action of the operating notation and mechanism of the engine...' She goes on to give a specific example: 'Supposing, for instance, that the fundamental relations of pitched sounds in the science of harmony and of musical composition were susceptible of such expression and adaptations, the engine might compose elaborate and scientific pieces of music of any degree of complexity or extent.'

16 Recognising the unique position of the EDSAC in computing history, a team of enthusiasts set about building a replica in 2012. It has been a challenging task, since no complete plan and only a handful of original components have survived. Yet, by piecing together overall logic diagrams, photographic evidence and the engineers' notebooks from the 1940s, the replica is now (as of early 2018) nearly complete. Remarkably, the thousands of vacuum tubes used in the EDSAC reconstruction are 1940s originals, recovered from government stockpiles.

17 The basic structure of a vacuum tube, first invented by John Ambrose Fleming in 1904 and then improved by Lee de Forest a few years later, superficially resembles that of an incandescent light bulb. As the name implies, the electrodes within the device exist in a vacuum. In most forms of the vacuum tube, the flow of electrons between the electrodes is controlled by thermionic emission, which means that current can only leap across the vacuum when one terminal, usually the cathode, is heated to a high temperature. By adjusting the arrangement and properties of the different electrodes, vacuum tubes control the flow of electrons across the vacuum. Current becomes unidirectional and vacuum tubes can be tailored to act as switches, amplifiers, oscillators, rectifiers (which convert alternating to direct current) or, in the case of the cathode ray tube, components of visual displays. The development of vacuum tubes with these different properties facilitated the invention of radar, radios, televisions, computers and many other electronic devices during the middle decades of the twentieth century.

18   Visionary mathematician and computer pioneer Alan Turing laid
      the theoretical foundations for modern computing in a 1936 research
      paper called 'On Computable Numbers, with an Application to the
      Entscheidungsproblem', which appeared in the *Proceedings of the
      London Mathematical Society*, Series 2, Volume 42, 1936–7, pp. 230–65.
      Turing described a theoretical machine, which executed mathematical
      calculations by manipulating a binary pattern of '1's and '0's on an
      infinitely long paper tape. He used this thought experiment to define
      the entire set of mathematical operations that can, in theory, be
      completed by a physical computer, which is what became known as a
      'Turing machine', as well as those operations that cannot be computed.
      He later extended his idea by describing a 'universal Turing machine'
      as any device that could be instructed to emulate the operation of any
      other Turing machine. It is this property, called 'Turing completeness',
      that early computer engineers strove to wire into their machines.
      'Turing complete' computers, also known as 'universal machines'
      can, in principle (and given the limitations of their finite memories),
      execute any and every calculation that is theoretically computable. This
      powerful concept is what makes today's computers so hugely adaptable
      and powerful.

19   In addition to these early computers, a wide range of mechanical
      calculators was invented during the first half of the twentieth century.
      In contrast to Babbage's automobile-sized Difference Engine, some,
      such as Curt Herzstark's 'Curta', fitted in the palm of a hand. Most of
      these devices, however, were restricted to executing basic arithmetic
      operations rather than flexible computing procedures. However, some
      of these mechanical calculators did push capabilities further. The
      differential analyser, invented by Vannevar Bush in 1931, could solve
      certain classes of differential equation. These sophisticated mechanisms
      proved crucial for studying dynamic systems such as bomb trajectories
      and the flow of electricity through the first large-scale power grids,
      for example. The Colossus computer, built by Tommy Flowers at the
      Post Office Research Station at Dollis Hill in north-west London in
      1943, was more flexible and powerful. Even so, its exclusive application
      was in solving cyphers. The ENIAC, designed and built by the engineers
      John Mauchly and J. Presper Eckert at the University of Pennsylvania
      in 1946, was the first general-purpose electronic computer. It could

be re-programmed, but doing so involved a great deal of technical working, changing switches and re-wiring. It often took several days to change the program. As a result, ENIAC's utility was restricted to computing ballistics tables and performing calculations for the development of nuclear weapons.

20 EDSAC was designed specifically with the end user in mind. As he explained in a 2010 interview, project leader Maurice Wilkes built it not to advance computer science but to help people answer real-world questions: 'When I started building the EDSAC, I had no doubt about who were going to be our users, they were people like myself ... I used to be doing things that would take perhaps a week solving ... with a desk machine, well that could be done rapidly with a digital computer. And those were the people who we made the EDSAC available to.' Maurice Wilkes' oral history interview, available online at http://www. bl.uk/voices-of-science/interviewees/Maurice-Wilkes. Wilkes foresaw that computers would become ubiquitous tools, accessible to everyone, which is why he decided to build EDSAC as a democratic resource for his university. Soon after the success of EDSAC, J. Lyons & Co., a British food and restaurant conglomerate that had supported the project, commissioned Wilkes' team to build a similar computer to help them run their business. LEO I (Lyons Electronic Office I) was the first computer designed explicitly for a business setting. It proved a success – soon other companies were placing orders for their own computers and Lyons became one of the world's first computing companies.

21 These Nobel laureates – molecular biologists John Kendrew and Max Perutz, physiologist Andrew Huxley and radio astronomer Martin Ryle – were not involved in building EDSAC, but they were early users of the computer who relied on EDSAC to analyse the complicated raw data they were gathering in their labs.

22 Just getting early computers like EDSAC to run a program was no mean feat. Computer science journalist Brian Hayes wrote: 'I am impressed that any useful work at all was ever done on the computer. The pioneer programmers who mastered the stern and unyielding machine were obviously prodigies of concentration and patience.' Quote from B. Hayes, 'The Discovery of Debugging', The Sciences, 33(4), 1993, pp. 10–13.

23 EDSAC, like all digital computers, ultimately deals in a binary pattern of ON/OFF, YES/NO, High/Low or 1/0 signals. Groups of binary digits, or 'bits', convey different pieces of information to a computer. For example, a string of 5 bits can exist in thirty-two different states (0,0,0,0,0 to 1,1,1,1,1 and the thirty different combinations of 1's and 0's in between). Each of these thirty-two binary states could represent a different letter or number, for example. By linking together different sequences of these strings of binary building blocks, increasingly complex instructions can be encoded. Commands for machines to store and manipulate data and to perform arithmetic operations upon it can be gradually constructed. This means that any system of symbols – all languages, all of mathematical notation, all logical operations, all musical notes, all images and videos – can be broken down into a binary pattern of bits. When it comes to solving problems, these can also be, in almost every case, reduced to a series of Yes/No questions – this gets to the essence of computation and, importantly, differentiates it from calculation, which requires only that machines complete sequential tasks in a linear and inflexible way. Another advantage of binary systems is that they avoid 'shades of grey', which means that their output is easier to interpret than analogue signals. They are also more robust in the face of noise and information can be copied endlessly and, in theory at least, flawlessly. The tedium of decomposing everything down to these endless strings of Yes/No signals is more than compensated for by the vast speed at which computers can churn through their binary code input. The latest microprocessors process sixty-four streams of binary information in parallel, completing hundreds of millions of instructions per second. The EDSAC, by contrast, had to loop through all its bits of information one at a time, in serial and executed just 650 instructions per second. That still greatly out-performed the fastest humans, even if they were assisted by mechanical calculators.

24 John Wilkins, *Mercury, Or, The Secret and Swift Messenger: Shewing how a Man May with Privacy and Speed Communicate His Thoughts to a Friend at Any Distance; Together with an Abstract of Dr. Wilkins's Essays Towards a Real Character and a Philosophical Language. Volume 6*, John Benjamins Publishing, 1708.

25 As quoted in James Gleick, *The Information: A History, A Theory, A Flood*, Pantheon Books, New York, 2011, p. 161.

26 To pose a problem on EDSAC you had to break your bigger question down into a series of basic mathematical operations, which had to be translated into a pattern of punch-marks on long strips of paper that were then fed into the computer. The only tangible output was the program's results, which eventually appeared as lines of text that emerged minutes or hours later on a teleprinter nearby.

27 The vacuum tubes and mercury delay lines that were used in EDSAC, and all the other computers of the time, are notoriously unreliable components. Just like the incandescent light bulbs that vacuum tubes superficially resemble, they generate a large amount of heat and their metal filaments are prone to burn out. The mercury delay lines, meanwhile, had a tiny memory capacity and were exquisitely sensitive to interference. It would be many years before computers evolved into the highly reliable systems that the modern world relies so heavily upon today.

28 Noyce's other masterstroke was to encapsulate the whole device within a layer of protective, insulating silicon oxide. Typical of the generally collaborative ethos of mid-twentieth-century computer engineering, Kilby and Noyce each generously deferred to the other as the originator of this new integrated circuit.

29 In his previous position at Fairchild Semiconductors, Faggin had helped to push forward transistor and integrated circuit technology. He'd previously built entire computers while working at Olivetti in Italy during the early 1960s.

30 In a 2004 interview, Faggin vividly described the moment when his first working chip arrived on his desk from the production plant one day in January 1971. It arrived at 6pm, just as most employees were heading home. Eager to see whether his new creation worked as planned, Faggin worked late into the night testing it. He'd been through the same agony three weeks previously, when a production error meant that he couldn't get any signal out of a first set of chips. This time everything worked. Faggin described it as 'one of the most elating moments of my life'. He returned home after 3am, waking his wife up with the cry, 'It works!' Federico Faggin Oral History, The Computer History Museum, Los Altos, 2004.

31 In a shrewd business move, Intel had offered Busicom a cheaper price if they agreed that Intel could maintain the intellectual and marketing rights to the innovative chip design.

32 Hopper observes that games have always had a role in pushing forward computing capabilities. Hopper's own PhD supervisor David Wheeler, who worked on the original EDSAC build, remembers creating and playing a game of noughts and crosses on the computer, as a way to push its performance. 'Games have always really stressed the system and continue to do so today,' explains Hopper.

33 Krzanich resigned from his position as CEO in June 2018.

34 In 2017 the International Technology Roadmap for Semiconductors was succeeded by the International Roadmap for Devices and Systems.

35 This is one of the properties of the quantum mechanical understanding of the world developed by leading physicists including Albert Einstein, Niels Bohr and Max Planck during the early twentieth century. Elementary particles such as quarks and leptons (electrons are a type of lepton) can behave as either particles or as waves, depending on how we measure them. They can also exist in more than one place at a time and they can spin in both directions at the same time and display the property of quantum entanglement, a phenomenon that Einstein famously branded 'spooky action at a distance'.

36 Many physical implementations of the qubit are currently being developed in research facilities around the world. Some use microchip-scale electrical circuits made from superconductors, where current can flow both clockwise and counter-clockwise at the same time. Others use the spin of atomic nuclei or electrons, which can be either 'up', 'down' or both, to represent qubits. There are also systems that trap single photons, which can exhibit either horizontal or vertical polarisation, or some combination of these states. Several other schemes are in the works too, and it remains to be seen which qubit design will prove most stable and sufficiently easy to implement in a practical quantum computer.

37 Furber went on to point out that encryption-coding experts have already devised encryption algorithms that cannot be cracked with a quantum computer. It is also possible to use quantum encryption. But as Furber sees it, 'all of this means you spend a lot more money to get back to where you started'.

38 Furber, working in close collaboration with Sophie Wilson at Acorn, built the first commercial microprocessor that implemented so-called 'Reduced Instruction Set Computing' (RISC), architecture, which had

been proposed and prototyped by academic researchers at Berkeley and Stanford. This was a new direction for semiconductor fabrication, which emphasised reduced size and complexity in order to optimise efficiency, rather than continuing the trend towards ever more transistors connected in increasingly complex ways. The efficiency and low power demand of RISC-based chips made them ideally suited for mobile and low-power devices and also for devices such as super-computers that must perform such intensive computational operations that power consumption and heat generation by traditional microprocessors had become a limiting factor. The Acorn RISC Machine led to the ARM Corporation, which now designs the chips used in the vast majority of smartphones, tablets and 'Internet of Things' devices used today. In 2017, the 100 billionth chip derived from Furber and Wilson's design was shipped.

39 Miniwatts Marketing Group reported that over 4.1 billion people, the equivalent of 54 per cent of the world population, accessed the Internet during 2017. Available online at https://www.internetworldstats.com/stats.htm

40 E. F. Codd, 'A Relational Model of Data for Large Shared Data Banks', *Communications of the ACM*, 13(6), 1970, pp. 377–87.

41 His relational database model was based on mathematical set theory. Data, Codd argued, should be stored in clearly cross-referenced tables, which allowed information to be stored, retrieved and queried easily and unambiguously. He also devised a simple language, which later evolved into the industry standard Structured Query Language (SQL), for users to manage their data with straightforward commands. With relational databases Codd did for data storage what Melvin Dewey's Decimal System had done for libraries a century before, though when it came to organising computer data, the stakes were higher – the modern world was fast becoming utterly reliant on databases.

42 At the time, IBM were heavily invested in an alternative system, based on complex data hierarchies, which they had first developed in the 1960s to manage the logistics and vast parts inventories for the Apollo space programme.

43 For a more detailed explanation of the impact of computerisation of legal discovery see James Bessen, 'The Automation Paradox', *Atlantic*, 9 January 2016.

44 S. Nakamoto. 'Bitcoin: A Peer-to-Peer Electronic Cash System', Cryptography mailing list at https://metzdowd.com, 2008.

45 To illustrate his point, Cerf describes how the 'wallets' that store cryptocurrencies like Bitcoin are vulnerable to hackers, even if the ledger that records and validates transactions is not. Regarding their energy consumption, a recent study calculated that mining Bitcoin could soon consumer as much as 2 per cent of the world's electricity production. See A. de Vries, 'Bitcoin's Growing Energy Problem', *Joule*, 2(5), 2018, pp. 801–5.

46 https://www.fourmilab.ch/babbage/sketch.html

47 J. F. Muratore, T. A. Heindel, T. B. Murphy, A. N. Rasmussen and R. Z. McFarland, 'Applications of Artificial Intelligence to Space Shuttle Mission Control', *The Proceedings of the Conference on Innovative Applications of Artificial Intelligence*, March 1989, pp. 15–22.

48 A. M. Turing, 'Computing Machinery and Intelligence', *Mind*, 49, 1950, pp. 433–60.

49 Go is very much more complicated than chess – there are, in fact, many more possible moves in a game of Go than there are atoms in the known universe. The hard-coded algorithms that use decision trees to allow computers to play chess unbeatably are wholly insufficient for Go, so more complex and flexible strategies are needed. Go grandmasters describe playing the game on the basis of instinct rather than logic. It seems that AlphaGo may, therefore, have approximated something similar to an instinct for the best move.

50 In 2017 a subsequent incarnation of this algorithm, called AlphaGo Zero, did not even need to be taught the rules of the game – it mastered Go from scratch and developed even more formidable Go-playing skills than its predecessor.

51 Artificial neural networks were first mooted as far back as the 1940s and are now one of the most potent approaches to machine learning. The inspiration behind them comes directly from neurobiologists' studies of the brain. For decades, neural networks failed to live up to the aspirations of their inventors. Then, at around the turn of the millennium, new algorithmic developments combined with the vast processing power of the latest computers, and neural networks gradually started to work and to outperform more traditional hard-coded machine learning strategies. Much of this turnaround stems

from an AI researcher with a background training in psychology and neuroscience called Geoff Hinton, who now works for Google.

52 The Go-playing AlphaGo algorithm used a combination of deep neural network-based and reinforcement-based learning to rapidly improve its playing prowess.

53 W. S. McCulloch and W. Pitts, 'A logical calculus of the ideas immanent in nervous activity', *The Bulletin of Mathematical Biophysics*, 5(4), 1943, pp. 115–33. Additionally, the 'backpropagation' algorithm that is today considered key to the utility of artificial neural networks was first conceived in 1960 (see J. Schmidhuber, 'Deep learning in neural networks: An overview', *Neural Networks*, 61, 2015, pp. 85–117).

54 Will Dunn, 'AI on the NHS', *New Statesman*, 31 January 2017.

55 NASA's Jet Propulsion Laboratory oversees most of their unmanned space missions. It develops and applies the AI algorithms that control autonomous spacecraft and planetary vehicles. As CEO of Beyond Limits AJ Abdallat explains, 'You need intelligence to do certain functions. The signal from space back to Earth takes a long time. We cannot operate these spacecraft remotely. We have to build in intelligence on-board these robotic missions. At Beyond Limits our mandate was to take those capabilities, add to them, enhance them, harden them for industrial applications and apply them to applications here on earth.

56 Hopper did not invent this example. There have been claims that an algorithm widely used in the US prison service to make criminal justice decisions contains an implicit racial bias against African Americans. See https://www.propublica.org/article/machine-bias-risk-assessments-in-criminal-sentencing

57 US Equal Employment Opportunity Commission Special Report: 'Diversity in High Tech'. Washington, DC, 2016. Available online at https://www.eeoc.gov/eeoc/statistics/reports/hightech

58 Ghahramani leads a team working on the 'Automatic Statistician' project (see https://www.automaticstatistician.com), which he calls 'the opposite of a black box machine-learning system. Something which is really as transparent a box as possible.' It is a machine-learning system that interprets data, finds patterns in the data and then it writes a report, in English, explaining what it has found: 'Basically it is trying to emulate a human data analyst or statistician.'

59 P. A. Mueller and D. M. Oppenheimer, 'The Pen is Mightier than the Keyboard: Advantages of Longhand over Laptop Note Taking', *Psychological Science*, 25(6), 2014, pp. 1159–1168.

60 Even today's most advanced computer styluses, such as the Apple Pencil, seek to replicate, rather than reinvent, the experience of using a pen or pencil.

61 This means that the first Cristals were almost exact contemporaries of EDSAC and the first generation of electronic computers. As computers matured and spread through societies, so did affordable pens.

4 CONNECT

1 Studies show that computer analysis of the list of web pages that people have 'liked' on Facebook is sufficient to predict personality traits, sexual orientation, political views intelligence, happiness, use of addictive substances and many other personal characteristics with startling precision. See W. Youyou, M. Kosinski and D. Stillwell, 'Computer-based personality judgements are more accurate than those made by humans', *Proceedings of the National Academy of Sciences*, 112(4), 2015, pp. 1036–40. And M. Kosinski, D. Stillwell and T. Graepel, 'Private traits and attributes are predictable from digital records of human behavior', *Proceedings of the National Academy of Sciences*, 110(15), 2013, pp. 5802–5.

2 You can read more on the DARPA website, https://www.darpa.mil/program/restoring-active-memory

3 Reed's Law proposes that network value increases as a function that is approximately 2^n, where n is the number of nodes in a network. From D. P. Reed, 'The Law of the Pack', *Harvard Business Review*, February 2001. It should be noted that neither Metcalfe's nor Reed's 'law' are true scientific laws, in the sense that neither distils a body of evidence. Neither has been verified as correct, presumably due to the poorly defined notion of network 'value' and the differing properties of different types of communication networks.

4 Nicholas Carr, *Utopia Is Creepy: And Other Provocations*, W. W. Norton & Company, New York, 2016

5 D. L. Hoffmann, C. D. Standish, M. García-Diez, P. B. Pettitt, J. A. Milton, J. Zilhão and M. Lorblanchet, 'U-Th dating of carbonate crusts

reveals Neandertal origin of Iberian cave art', *Science*, 359(6378), 2018, pp. 912–15.

6   In AD 828, Muhammad ibn Musa Al-Khwarizmi wrote a treatise called *Al-Kitab al-mukhtasar fi hisab al-jabr wa'l-muqabala*, usually referred to as *al-Jabr*, which translates as *On Completion*. Al-Khwarizmi made mathematics easier and more powerful than ever before, and his principles spread quickly across the Islamic world. Then, during the thirteenth century, the Italian Leonardo of Pisa, better known by his nickname 'Fibonacci', discovered the excitement of Middle-Eastern mathematics and imported the key elements, including the digits, to Europe. Although eagerly adopted by mathematicians, the new system diffused into everyday business and spread across the continent surprisingly slowly – although it was less convenient for calculations, the incumbent system of Roman numerals worked. In Florence, the use of Arabic numerals was forbidden for finance and book-keeping. It was not until the invention of reliable printing presses that the Arabic numerals and the decimal system would assert their dominance on Europe and the rest of the world.

7   Florian Cajori, *A History of Mathematics*, American Mathematical Society, 1991.

8   Movable-type printing had been tried before, but no one had made it into a sufficiently reliable and versatile technique. In China, wooden and porcelain printing blocks had been used since the second century BC, and metal typefaces were used to print Zen Buddhist literature in fourteenth-century Korea. In Europe too, laboriously carved wood blocks were used to print some simple texts, but these early efforts were slow and inflexible, and none of them were widely adopted.

9   According to some sources, the first Gutenburg Bibles cost as much as thirty florins, roughly three times the annual salary of a clerk. (See Lesley B. Cormack and Andrew Ede, *A History of Science in Society: From Philosophy to Utility*, Broadview Press, 2004, p. 95.) However, in 1468 Giovanni Andrea Bussi, Bishop of Aléria, wrote to the Pope, saying, 'In our time God gave Christendom a gift which enables even the pauper to acquire books. Prices of books have decreased by eighty percent.' (R. Hirsch, *Printing, selling and reading, 1450–1550*. Harrassowitz, 1974)

10   To put this rate of technological uptake in perspective, the first deployment of the Internet took place over a similar timescale. As we will see, the ARPANET, the forerunner of today's Internet, was conceived by engineers working for the US Department of Defense Advanced Research Projects Agency (DARPA) during the 1960s. It did not change into the Internet and expand out of academic and military institutions and into people's homes until the 1990s. In 1993, only 130 websites existed; today there are well over a billion.

11   Filippo de Strata, *Polemic against Printing* (translated by Shelagh Grier and edited by Martin Lowry), Hayloft Press, Birmingham, 1986.

12   Queen Victoria wrote the first message. Predictably, it was less hyperbolic than that of the US President. She hoped the telegraph cable would provide 'an additional link between the nations whose friendship is founded on their common interest and reciprocal esteem'. These idealistic proclamations about the telegraph network from both sides of the Atlantic were, at best, premature. That first wire across the Atlantic failed less than three weeks after it carried its first message. It would be another eight years before Isambard Kingdom Brunel's huge steam ship, the SS *Great Eastern*, laid a more durable and effective cable across the ocean. The telegraph ushered us into the age of telecommunications, but it was not the route to peace and enlightenment that some had hoped it would be. From the American Civil War to the First World War, the decades that followed its adoption were far from peaceful, as wired, and then wireless, telegraphy was eagerly adopted by the military to orchestrate their campaigns.

13   These calls are not, of course, free. Prosaically, they depend on having access to computer hardware and subscriptions to Internet service providers, for example. Less tangibly, there is usually a less visible cost, since data about how, when and why people use these 'free' calling services are a source of value for the service provider – by providing information for the targeting of advertisements, for example.

14   Miniwatts Marketing Group report that over 4.1 billion people, or the equivalent of 54 per cent of the world population, accessed the Internet during 2017. Available online at https://www.internetworldstats.com/stats.htm

15   The Internet and World Wide Web were founded according to the principle of 'net neutrality', the principle that telecommunications

operators must treat all data equally, rather than allowing preferential access to content from different sources or with differential pricing. In some parts of the world this principle has been upheld robustly, while in other countries, including China, Turkey, Russia and Iran, government and corporate censorship and differential access controls are reported to be more common, limiting connectivity and restricting access to certain content.

16 'Email Statistics Report, 2017–2021', Radicati Group Inc., Palo Alto, 2017. Available online at http://www.radicati.com/wp/wp-content/uploads/2017/01/Email-Statistics-Report-2017-2021-Executive-Summary.pdf

17 Darya Gudkova, Maria Vergelis, Tatyana Shcherbakova and Nadezhda Demidova, 'Spam and phishing in Q3 2017', *Kaspersky Lab*, 2017. Available online at https://securelist.com/spam-and-phishing-in-q3-2017/82901/

18 John Browne, Robin Nuttall and Tommy Stadlen, *Connect: How Companies Succeed by Engaging Radically with Society*, Random House, London, 2016.

19 D. Weitzner, 'Learning from Tim Berners-Lee on the occasion of receiving the Turing Award', *MIT Internet Policy Research Initiative*, 4 April 2017. Available online at https://internetpolicy.mit.edu/timbl-turing-lessons/

20 J. P. Barlow, 'A Declaration of the Independence of Cyberspace', *The Humanist*, 56(3), 1996, p. 18.

21 B. Gates, 'The Internet Tidal Wave', Internal Microsoft Memo, 26 May 1995. Available online at https://www.justice.gov/sites/default/files/atr/legacy/2006/03/03/20.pdf

22 N. Negroponte, 'Internet is way to world peace', *SciTech*, 25 November 1997.

23 C. Stoll, 'Why the Web won't be Nirvana', *Newsweek*, 26 February 1995.

24 Eli Pariser, *The Filter Bubble: What the Internet is Hiding from You*, Penguin, London, 2011.

25 The quote is from Francis Bacon's 1620 philosophical work *Novum Organum Scientiarum*, in *Collected works of Francis Bacon*, Routledge, Oxford, 1996.

26 In sum, fake news stories stimulated more engagement with Facebook users than verifiable ones did during the last three months of the 2016 US presidential campaign. From C. Silverman, 'This Analysis Shows

How Viral Fake Election News Stories Outperformed Real News On Facebook', *BuzzFeed*, 16 November 2016.

27 S. Vosoughi, D. Roy and S. Aral, 'The spread of true and false news online', *Science*, 359(6380), 26 May 1995, pp. 1146–51.

28 R. Meyer, 'The Grim Conclusions of the Largest-Ever Study of Fake News', *Atlantic*, 8 March 2018.

29 Looking ahead, the challenges of distinguishing truth from falsehood are likely to grow. Feed a twenty-minute recording of someone's voice to a new computer program and you can make that person say exactly what you want them to say. Hollywood studios, meanwhile, have begun populating their films with digital actors that are impossible to distinguish from their flesh and blood co-performers. These tools will become more convincing, cheaper and easier to use.

30 R. Edelman, 'An Implosion of Trust', *Edelman*, 8 March 2017. Available online at http://www.edelman.com/p/6-a-m/an-implosion-of-trust. Also the Edelman Trust Barometer, 2017.

31 S. Athey, C. Catalini and C. Tucker, *The Digital Privacy Paradox: Small Money, Small Costs, Small Talk, National Bureau of Economic Research*, 2017.

32 C. Kang and S. Frenkel, 'Facebook Says Cambridge Analytica Harvested Data of Up to 87 Million Users', *New York Times,* 4 April 2018.

33 The 'Solid' programme at MIT aims to 'radically change the way Web applications work today, resulting in true data ownership as well as improved privacy.' See https://solid.mit.edu

34 R. Botsman, 'Big data meets Big Brother as China moves to rate its citizens', *Wired*, 28 November 2017.

35 B. Scheier, 'The risks – and benefits – of letting algorithms judge us', CNN.com, January 2016.

36 A. Blair, 'Reading Strategies for Coping with Information Overload *ca.* 1550–1700', *Journal of the History of Ideas*, 64(1), 2003, pp. 11–28.

37 'There is an important debate to be had about screen time, but we need quality research and evidence to support it', *Guardian*, 6 January 2017.

38 There is evidence that millennials are reading more books than older generations. See A. Perrin, 'Book Reading 2016', *Pew Research Center Report*, 1 September 2016.

39 'Tech Control of Your Future on Social Media', Headmasters' & Headmistresses' Conference (HMC), UK. October 2017. Available

online at https://www.hmc.org.uk/wp-content/uploads/2017/10/Tech-Control-Survey.pdf

40 'Paying for News: Why People Subscribe and What It Says About the Future of Journalism', The Media Insight Project, a collaboration between the American Press Institute and The AP-NORC Center, May 2017. Available online at https://www.americanpressinstitute.org/publications/reports/survey-research/paying-for-news. Also L. Graves and F. Cherubini, 'The rise of fact-checking sites in Europe', *Reuters Institute for the Study of Journalism*, 2016. Available online at http://reutersinstitute.politics.ox.ac.uk/our-research/rise-fact-checking-sites-europe

41 In per capita terms, the difference is small and shrinking – in both Africa and Europe there is, on average, more than one mobile phone contract per person. Data from the World Bank.

42 J. Masinde, 'Kenya's M-Pesa platform is so successful regulators worry it could disrupt the economy', *Quartz*, 28 December 2016.

43 To illustrate his point, Juma explained how smartphone cameras are saving and improving lives in remote areas – medics and public health officials already use them to diagnose conditions and offer the best guidance remotely. Smartphones can measure vital signs, screen for eye health and deliver the very latest research to the most isolated regions.

44 It is not easy to establish precisely how much of our communication depends on non-verbal cues. Claims of this kind can usually be traced back to the work of a psychologist called Albert Mehrabian in the 1960s. Mehrabian and his collaborators tested the way inconsistencies between facial expression, tone of voice and the verbal content of short, spoken messages affected people's ability to interpret the preferences of the communicator – the conclusion was that expression and tone were more powerful cues than the precise words spoken.

45 'There's a lot of baggage in human relationships too,' Breazeal continued. We can design robots that people feel supported by and never feel embarrassed or inhibited in front of; 'You can ask [a robot] the same question twenty times and not worry about feeling stupid or wasting somebody's time,' she says.

46 G. Lewis-Kraus, 'The Great AI Awakening', *New York Times Magazine*, 14 December 2016, pp. 1–37.

47 Further illustrating his point about the way computer systems are getting better at comprehending us, Shemtov described how online searches fifteen years ago relied on matching single keywords. Now, although we can type out whole questions and, if we are lucky, get meaningful answers, seamless dialogue with computers is still impossible.

48 P. Gosh, '£1m Queen Elizabeth Prize: Digital camera tech lauded', *BBC News*, 1 February 2017.

49 Fossum reported how, in a 1995 interview for *Businessweek*, he mentioned that 'portable video phones' would be a possible application of the new technology. 'Little did I know that this would, in fact, be the "killer app"', he said.

50 'I didn't know anything about entrepreneurship when we started this company', Fossum says, explaining how he quickly learned that 'luck and timing is just an incredible part of a successful start-up'. That, he appreciated, could be frustrating because, 'as engineers ... we're used to controlling everything'.

51 Y. Wang and M. Kosinski, 'Deep neural networks are more accurate than humans at detecting sexual orientation from facial images', *Journal of Personality and Social Psychology*, 114(2), 2018, pp. 246–57. When people were asked to categorise facial images according to sexuality, they achieved a lower accuracy than the AI algorithm: 61 per cent for men and 54 per cent for women, versus 81 per cent and 74 per cent respectively for the algorithm.

52 It is quite conceivable that today's most popular social media platforms, including Facebook, Instagram, Snapchat and Twitter, could be made obsolete by new innovations during the next decade or even sooner.

5 BUILD

1 To make their cement, the Mayans built circular towers of wood, ten metres wide and ten metres high. After piling limestone and clay-containing minerals on top, they lit a fire at the base of a vertical channel at the centre of the temporary kiln. As the wood burned, oxygen was sucked through this channel, fanning the flames. The temperature at the core of these makeshift blast furnaces reached up to 1,600 degrees Celsius, hot enough to melt the limestone and convert it into a chemical called tricalcium silicate (Ca_3SiO_5), commonly known as alite, the key chemical component of hydraulic cement. As

described by James O'Kon in *The Lost Secrets of Maya Technology*, New Page, Newburyport, 2012.

2 The Romans also used hydraulic cement, at least as far back as 25 BC, when Vitruvius described various mortars in his *Ten Books of Architecture*. But they took a different route to cement production, one that demanded less intense heat. Roman builders mixed quicklime (made from limestone in a kiln at about 900 degrees Celsius) with a form of volcanic ash called *pozzolana*. This process meant that alite could be produced at a lower temperature than the direct route taken by the Maya.

3 For reasons that archaeologists still cannot fully decipher, the planet Venus had particular significance. The Mayans used their astronomical knowledge to devise calendrical systems that recorded the passage of time with a complexity and level of precision that exceeds anything developed by European or Asian cultures of the same period.

4 The most comprehensive and readable summary of the factors leading to the demise of the Mayan civilisation is in *Collapse: How Societies Choose to Fail or Succeed* by Jared Diamond, Penguin, New York, 2005.

5 The following doom-laden quote from Malthus describes the 'Malthusian catastrophes' that he predicted: 'Famine seems to be the last, the most dreadful resource of nature. The power of population is so superior to the power of the Earth to produce subsistence for man, that premature death must in some shape or other visit the human race. The vices of mankind are active and able ministers of depopulation. They are the precursors in the great army of destruction, and often finish the dreadful work themselves. But should they fail in this war of extermination, sickly seasons, epidemics, pestilence, and plague advance in terrific array, and sweep off their thousands and tens of thousands. Should success be still incomplete, gigantic inevitable famine stalks in the rear, and with one mighty blow levels the population with the food of the world.' From Thomas Malthus, *An Essay on the Principle of Population*, J. Johnson, London, 1798, p. 61.

6 Edward Fitzgerald, *Rubaiyat of Omar Khayyam*, Oxford University Press, Oxford, 2009.

7 The ambition to build imposing domes created many serious engineering challenges. The dome of Florence's cathedral, for example, remained unfinished for many years because no one could resolve the outward thrust of the dome on the supporting structures below.

Eventually a competition was held, which was won by Brunelleschi – his solution was to incorporate a great hoop of timber and iron into the base of the dome. For the dome of St Paul's, Wren created a channel in the stone structure, into which he laid two great sets of iron chains, which were tightened before the channel was filled with lead. The dome of St Paul's is in fact a triple dome made of three layers, each with different structural and decorative requirements.

8   The early and extensive flow of inspiration and technical knowhow between East and West is plain to see for anyone visiting the great churches and mosques of the Renaissance period. Architectural historians think it is almost certain that influence must have flowed in both directions between Mimar Sinan, the great architect of the Ottoman Empire, and his Italian contemporaries Michelangelo and Andrea Palladio, for example. In his second tract on architecture, Wren explained that for the ingenious triple-dome construction at St Paul's he followed the technique used at the Hagia Sophia mosque in Istanbul. From Helen Hills (ed.), *Rethinking the Baroque*, Ashgate Publishing, Oxford, 2011.

9   L. M. Bettencourt, J. Lobo, D. Helbing, C. Kühnert and G. B. West, 'Growth, innovation, scaling, and the pace of life in cities', *Proceedings of the National Academy of Sciences*, 104(17), 2007, pp. 7301–6.

10   Urban population data from 'The United Nations Population Division's World Urbanization Prospects report', via the World Bank. Available online at https://data.worldbank.org/indicator/SP.URB.TOTL.IN.ZS

11   'World Urbanization Prospects: The 2018 Revision', Department of Economic and Social Affairs. Available online at https://esa.un.org/unpd/wup

12   R. Dobbs, J. Remes, J. Manyika, C. Roxburgh, S. Smit and F. Schaer, *Urban world: Cities and the Rise of the Consuming Class*, McKinsey Global Institute, 2012.

13   It is a great mystery of architectural history but the art of making cement, which as we have already seen was widespread in both ancient Greece and Rome and independently invented by the Maya, was almost entirely lost in Europe for the millennium and a half that followed the fall of the Roman Empire.

14   David Deming, *Science and Technology in World History, Volume 2: Early Christianity, the Rise of Islam and the Middle Ages*, McFarland, Jefferson, 2012.

15 Quotes by Neave Brown are from a recent interview: Jessica Mairs, 'High-rise buildings should only house the very rich, says Neave Brown', *Dezeen*, 5 October 2017.

16 M. Kothari, 'Economic, social and cultural rights, Report of the Special Rapporteur on adequate housing as a component of the right to an adequate standard of living', Commission on Human Rights, Geneva. And J. Woetzel, S. Ram, J. Mischke, N. Garemo and S. Sankhe, 'A blueprint for addressing the global affordable housing challenge', *McKinsey Global Institute*, 2014.

17 Aravena explained that the figure of $10,000 per family home is the average value of funds made available by housing ministries and loans made by organisations such as the World Bank and the International Monetary Fund.

18 'An earthquake shows that Mexico has learned from past disasters', *Economist*, 9 September 2017.

19 C. Deuskar, 'Is China's Pearl River Delta the world's biggest city?', *The World Bank Data Blog*, 2015. Also available at http://blogs.worldbank. org/opendata/new-data-and-research-help-measure-decade-urban-expansion-across-east-asia

20 Vaclav Smil, *Making the Modern World: Materials and Dematerialization*, John Wiley & Sons, New York, 2016.

21 'What China can learn from the Pearl River Delta', *Economist*, 8 April 2017.

22 K. Wan and J. Liu, 'The Spatiotemporal Trend of City Parks in Mainland China between 1981 and 2014: Implications for the Promotion of Leisure Time Physical Activity and Planning', *International Journal of Environmental Research and Public Health*, 14(10), 2017, p. 1150.

6 ENERGISE

1 The first 'humans', defined as members of the genus *Homo*, were members of the species *Homo habilis*, which evolved a little over 2 million years ago. Our own species, *Homo sapiens*, appeared between 200,000 and 300,000 years ago.

2 For an accessible account of the evidence for a pivotal role that fire played in our evolution, see Richard Wrangham, *Catching Fire: How Cooking Made Us Human*, Profile Books, London, 2010. For a recent

academic review see: J. A. J. Gowlett, 'The discovery of fire by humans: a long and convoluted process', *Philosophical Transactions of the Royal Society B*, 2016.

3 For readable summaries of the origins of agriculture, including the idea that it was a mixed blessing for many, see: Jared Diamond, *Guns, Germs and Steel: A Short History of Everybody for the Last 13,000 Years*, Random House, London 1998; Yuval Noah Harari, *Sapiens: A Brief History of Humankind*, Harper, New York, 2015; Vaclav Smil, *Energy and Civilization: A History*, MIT Press, Cambridge, 2017. There is some archaeological evidence that members of pre-agricultural societies were taller, fitter and had larger brains than members of later agrarian communities.

4 Skoglund describes how the clearest genetic signatures that are associated with the transition to a farming-based lifestyle are the persistence of expression of a gene that maintains lactose tolerance from infanthood into adulthood; and the expansion of number of copies of a gene called amylase, which is needed to break down starch. He also explains how Inuit people, whose technologies allowed them to survive in Arctic environments, are genetically adapted to be better able to digest animal fats, consistent with the high fat content of their diet. Clear evidence of evolutionary change in response to today's fast-changing technologically dependent lifestyle is much more challenging to detect reliably. A recent study showed that there was enrichment in the genomes of longer-lived people for genetic variants that provide resistance to dementia and make dependence on nicotine less likely, consistent with the idea that there has been recent selective pressure favouring these two traits. To read more on this topic, including further examples of genetic adaptations, past and present, see S. Marciniak and G. H. Perry, 'Harnessing ancient genomes to study the history of human adaptation', *Nature Reviews Genetics*, 18(11), 2017 and P. Skoglund, P. I. Mathieson, 'Ancient Human Genomics: The First Decade', *Annual Review of Genomics and Human Genetics*, 19, 2018 pp. 381–404.

5 Vannoccio Biringuccio, *The Pirotechnia of Vannoccio Biringuccio*. translated by Cyril Stanley Smith and Martha Teach Gnudi, Dover Publications, New York, 1990.

6   In parts of China, for example, coal was used to melt iron ore as far back as the Han dynasty in the second century BC.

7   Marco Polo. *The Travels of Marco Polo*, translated by Henry Yule and edited and annotated by Henri Cordier. John Murray, London, 1920.

8   Arnold Toynbee, *Lectures on the Industrial Revolution In England: Public Addresses, Notes and Other Fragments*, together with *A Short Memoir* by B. Jowett, Rivington's, London, 1884.

9   For real wages see G. Clark, 'The Condition of the Working Class in England, 1209–2004', *Journal of Political Economy*, 113(6), 2005, pp. 1307–40. For height trends see T. J. Hatton and B. E. Bray, 'Long run trends in the heights of European men, 19th–20th centuries', *Economics & Human Biology*, 8(3), pp. 405–13. For life expectancy see S. Szreter. and G. Mooney, 'Urbanization, Mortality, and the Standard of Living Debate: New Estimates of the Expectation of Life at Birth in Nineteenth-Century British Cities', *The Economic History Review*, 51(1), 1998, pp. 84–112. For an accessible summary, see https://www.economist. com/blogs/freeexchange/2013/09/economic-history-0

10 Emma Griffin, *Liberty's Dawn: A People's History of the Industrial Revolution*, Yale University Press, New Haven, 2014.

11 Ibid.

12 William Stanley Jevons, *The Coal Question; An Inquiry Concerning the Progress of the Nation, and the Probable Exhaustion of Our Coal Mines*, MacMillan, 1865.

13 Vaclav Smil, *Energy and Civilization: A History*, MIT Press, Cambridge, 2017, p. 177.

14 Wallace Kaufman, *No Turning Back: Dismantling the Fantasies of Environmental Thinking*, Basic Books, 1994.

15 *BP Statistical Review of World Energy*, 2017. Available online at http:// www.bp.com/en/global/corporate/energy-economics/statistical-review-of-world-energy/oil/oil-production.html

16 Vaclav Smil, *Enriching the Earth: Fritz Haber, Carl Bosch, and the Transformation of World Food Production*. MIT Press, Cambridge, 2004.

17 Fritz Haber also invented poison gases that were used in the First World War, as well as the process used to make the disinfectant Zyklon B that was used by the Nazis to kill over a million people in the Holocaust. The Churchill quote is from a Parliamentary debate on 25 April 1918,

in Richard Langworth (ed.), *Churchill by Himself: The Definitive Collection of Quotations*, PublicAffairs, New York, 2008, p. 469.

18 Most hydraulic fracturing fluids incorporate low concentrations of chemicals, to enhance the recovery of oil and gas. Although companies in the US are not required to release information about the chemicals they use, since it is classed as proprietary information, the most commonly used substances are thought to be methanol (an alcohol), hydrochloric acid and hydro-treated light petroleum distillates (a combination of hydrocarbons).

19 'Hydraulically fractured horizontal wells account for most new oil and natural gas wells', US Energy Information Administration, January 2018. Available online at https://www.eia.gov/todayinenergy/detail.php?id=34732

20 E. Barth-Naftilan, J. Sohng and J. E. Saiers, 'Methane in groundwater before, during, and after hydraulic fracturing of the Marcellus Shale', *Proceedings of the National Academy of Sciences*, 115(27), 2018, pp. 6970–6975.

21 Parson's showmanship paid off. Soon the Navy were eagerly building turbine-powered ships and the whole shipping industry was transformed. The vastly improved power-to-weight ratio meant that ships were not just faster – they could be built much larger. If it had been powered by a traditional steam engine, for example, the *Titanic* could not have carried enough coal to even attempt a crossing of the Atlantic.

22 A dynamo is an electricity generator that produces direct current. An immobile stator provides a constant magnetic field and contains coils of wire that rotate within the field to generate current, according to Faraday's law of induction. A commutator acts as a rotary switch to convert the alternating current generated by the rotating coils into direct current. Michael Faraday demonstrated the key principles in 1831–2. The first working dynamos found industrial applications during the 1860s.

23 Robert J. Gordon, *The Rise and Fall of American Growth: The U.S. Standard of Living since the Civil War*, Princeton University Press, 2016.

24 C. T. Perry and K. M. Morgan, 'Bleaching drives collapse in reef carbonate budgets and reef growth potential on southern Maldives reefs', *Scientific Reports*, 7, 2017.

25 This report from the US National Academies of Science describes many of the reported effects of climate change, to date. 'Our Changing Climate', US Global Change Research Program Washington, DC, 2014. Available online at http://nca2014.globalchange.gov/highlights/overview/overview

26 'Short-term energy outlook', US Energy Information Administration, April 2018. Available online at https://www.eia.gov/todayinenergy/detail.php?id=34872

27 A. S. Eddington, 'The Internal Constitution of the Stars', *Scientific Monthly*, 11(4), 1920, pp. 297–303.

28 In 1911 the physicist Lord Rutherford proposed that nuclear atoms comprised a heavy, positively charged nucleus, surrounded by negatively charged electrons. Shortly before Rutherford proposed his atomic model, in 1905, Albert Einstein deduced that mass and energy were equivalent. This he summarised in his famous equation $E=mc^2$ (energy is equal to mass multiplied by the speed of light squared). Even the tiny mass of the atomic nucleus must, therefore, contain a large amount of energy. A nuclear fusion reaction is designed to liberate some of this energy, by reacting to the nuclei of the hydrogen isotopes deuterium and tritium, to create a new nucleus of helium, whose net mass is less than the combined mass of the two isotopes of hydrogen. This loss in mass produces an enormous amount of energy – nuclear fusion reactions of this sort are the source of the energy produced by our sun and all other stars.

29 Practical fusion fuel is most likely to consist of two isotopes of hydrogen, deuterium and tritium – both of which can, in principle, be derived from seawater.

30 JET (the Joint European Torus) is a tokamak, a toroidal (doughnut-shaped) reactor that uses a powerful magnetic field to contain and add energy to nuclear fusion fuel, in the form of a cloud of plasma. The first tokamaks were built in the 1950s by Soviet physicists.

31 Plasma is a state of matter created by adding huge amounts of energy to a gas. In a plasma, many electrons become separated from their atomic nuclei, which creates a highly energetic mixture of positively charged ions and negatively charged electrons that is highly responsive to electromagnetic fields. As much as 99 per cent of the matter in the universe is thought to be in a plasma state.

32   For now, the most powerful fusion reaction ever achieved took place back in 1997, when JET produced sixteen megawatts of fusion power. That output came at a cost of twenty-four megawatts of grid-supplied electricity, a net loss of energy. Until the amount of energy harvested from fusion reactors far exceeds the energy needed to start the fusion reaction, this technology is of no practical value.

33   Turbulence occurs when the motion of flowing fluid (either a liquid, a gas or, as in nuclear fusion, a plasma) becomes chaotic and unpredictable. In fluids with relatively lower viscosity (say, water versus crude oil), flow is more likely to become turbulent. Particles within the fluid interact with each other in complex ways, creating unpredictable vortices and eddies. As a result, measurements of the velocity of a turbulent fluid will show continuous changes in both magnitude and scale at any given position. Richard Feynman said in the 1960s that 'Turbulence is the most important unsolved problem of classical physics,' and the phenomenon remains poorly understood today.

34   Tatsujiro Suzuki, 'Six years after Fukushima, much of Japan has lost faith in nuclear power', *The Conversation*, 9 March 2017.

35   International Energy Agency, 'World Energy Outlook: Organisation for Economic Co-operation and Development', 2017. Available online at https://www.iea.org/media/publications/weo/WEO2017ToC.pdf

36   This figure is reached by multiplying the capacity of the plant (6,800 kilowatts) x 24 hours x 365.25 days per year x the estimated solar capacity factor for Sicily (0.3, i.e. the portion of each day when solar power is generated), and then dividing by the average annual household energy consumption in Italy (2,432 kilowatt-hours, according to the World Energy Council). See https://wec-indicators.enerdata.net/household-electricity-use.html

37   'World's largest solar park launched in Karnataka', *Economic Times*, 1 March 2018.

38   There is a long history of engineers pointing to the vast potential of solar energy and imagining schemes that could harvest it on a grand scale. In 1913, for example, American engineer Frank Schuman outlined designs to cover a 20,000-square-mile patch of the Sahara Desert with a type of solar engine that he had designed – it would generate 270 million horsepower, which, as he noted, was equal to all the fuel burned around the world in 1909. Since then the world

has grown, and more energy is needed. More recently, the Land Art Generator calculated that, with today's photovoltaics, it would take around 190,000 square miles to meet the world's entire energy needs (see calculations at http://landartgenerator.org/blagi/archives/127, which are backed up by 'back of the envelope' calculations shared with me by Professor Richard Friend).

39 Solar cost data is from Bloomberg New Energy Finance, as summarised by Michael Liebrich in his 'State of the Industry' address at the Bloomberg New Energy Finance Summit in London, 2017. Available online at https://data.bloomberglp.com/bnef/sites/14/2017/09/BNEF-Summit-London-2017-Michael-Liebreich-State-of-the-Industry.pdf

40 International Energy Agency, 'Snapshot of Global PV Markets: Organisation for Economic Co-operation and Development', 2016. Available online at http://www.iea-pvps.org/fileadmin/dam/public/report/statistics/IEA-PVPS_-_A_Snapshot_of_Global_PV_-_1992-2016__1_.pdf

41 Alexandre-Edmond Becquerel discovered the photoelectric effect in 1839, when he showed that light could stimulate an electric current to flow between two brass plates that are immersed in a conductive liquid. When photons of light hit a semiconductor like silicon or lead halide perovskites, electrons are ejected from the semiconductor, causing an electrical current to flow.

42 One hope for perovskite solar panels is that they could also be produced in different shades, each one of which would harvest a different slice of sunlight. Cells that tap into every wavelength of light available could be engineered by layering different shades on top of one another. If this were successful, they would beat the overall efficiency of silicon cells, which only absorb at a single wavelength. Friend explained this effect: 'It doesn't matter whether it's an infrared photon of 1.1 eV that is just absorbed in silicon or a blue photon of 2.7 eV – they both make the same contribution to the solar cell. But in the case of the blue photon, more than half of its energy is dumped as heat, since it had too much energy. That so-called thermalisation loss is the principal reason why the very best silicon cells on sale today are only up to 25 percent efficient.'

43 One strategy that is being explored by Snaith and others who seek to commercialise perovskite solar cells is to build tandem cells that

layer it on top of silicon cells – this could be a way to use perovskites to improve the performance of the current best-proven photovoltaic technology without threatening the huge investment that has already been made in silicon cell production facilities.

44   R. Harrabin, 'China embarked on wind power frenzy, says IEA', *BBC News*, 20 September 2016.

45   Greenpeace, 'Accelerating the Energy Transition: The Co-benefits of Wind and Solar PV Power in China', 2017.

46   A recent report by Nigel Brandon's engineering colleagues at Imperial College describes the value of flexibility in the national grid. Imperial College London and NERA Economic Consulting, 'Value of Flexibility in a Decarbonised Grid and System Externalities of Low-Carbon Generation Technologies', 2015. Available online at https://www.theccc.org.uk/wp-content/uploads/2015/10/CCC_Externalities_report_Imperial_Final_21Oct20151.pdf

47   Lithium ion batteries store chemical potential energy at solid electrodes in the form of lithium ions, which move through an electrolyte from the battery's cathode (positive electrode) to the anode (negative electrode) during charging, and in the reverse direction during discharging. Vanadium flow batteries, by contrast, store chemical potential energy in the electrolyte, by using the fact that vanadium (V) can exist in four different oxidation states, or charges. During charging, energy is stored by oxidising $V4+$ to $V5+$ on the positive side of the cell, and reducing $V3+$ to $V2+$ on the negative side. During discharge, the process is reversed, creating a current that flows through the external circuit.

48   For an example of a recent step forward in this battery technology, see H. Yin, B. Chung, F. Chen, T. Ouchi, J. Zhao, N. Tanaka and D. R. Sadoway, 'Faradaically selective membrane for liquid metal displacement batteries', *Nature Energy*, 2018, p. 1.

49   Excess power can also be converted into potential energy. The best proven approach is to pump water uphill into a dam using surplus power and let it rush through hydroelectric turbines when the grid demands more power. Other systems put surplus power into liquefying nitrogen or compressing air and then turn it back into electricity as the physical process is reversed. A further idea is to use surplus power to spin carbon fibre flywheels at high speeds – unlike traditional batteries, they will suffer minimal depreciation with age and repeated

use and are a way to store or release power very quickly, best suited to balancing short-term peaks and troughs in grid load.

50 P. E. Dodds, I. Staffell, A. D. Hawkes, F. Li, P. Grünewald, W. McDowall and P. Ekins, 'Hydrogen and fuel cell technologies for heating: A review', *International Journal of Hydrogen Energy*, 40(5), 2015, pp. 2065–83

51 World Energy Council, 'World Energy Resources', 2016. Available online at https://www.worldenergy.org/publications/2016/world-energy-resources-2016/

52 The combined-cycle gas turbine power plant that opened in Bouchain, France, in 2016 achieves an efficiency of over 62 per cent, making it the most efficient gas-fuelled electricity generation facility ever built.

53 International Energy Agency, 'World Energy Outlook: Organisation for Economic Co-operation and Development', 2017. Available online at https://www.iea.org/media/publications/weo/WEO2017ToC.pdf

54 The ever-present risk is that growing plants to mitigate climate change and/or to provide an energy source for the production of biofuels would reduce the amount of land available for agriculture, an important consideration, as the world's population is expected to continue to grow for the next five decades. There has recently been some encouraging progress in adapting and improving enzyme systems from bacteria, which could be used to turn biomass that would otherwise go to waste, such as straw and chaff, into biofuel. This would be one way to combine agriculture and biofuel production on the same land, while also minimising waste and carbon dioxide emissions. For a review of this approach, see L. Artzi, E. A. Bayer and S. Moraïs, 'Cellulosomes: bacterial nanomachines for dismantling plant polysaccharides', *Nature Reviews Microbiology*, 15(2), 2017, p. 83.

55 Z. Chen, J. J. Concepcion, M. K. Brennaman, P. Kang, M. R. Norris, P. G. Hoertz and T. J. Meyer, 'Splitting CO_2 into CO and O_2 by a single catalyst', *Proceedings of the National Academy of Sciences*, 109(39), 2012, pp. 15606–11.

56 K. Bourzac, 'Liquid sunlight', *Proceedings of the National Academy of Sciences*, 113(17), 2016, pp. 4545–8 and a recent report from MIT, 'Solar to Fuels Conversion Technologies: A Working Paper', 2015.

57 International Energy Agency (IEA) and the Renewable Energy Agency (IRENA), 'Perspectives for the energy transition – investment for a low carbon energy system', 2017.

7 MOVE

1   Concorde's nose cone and leading edges heated up to as much as 127 degrees Celsius during flight, when heat caused the whole aircraft to expand by as much as thirty centimetres. An advanced alloy of aluminium and a formulation of white paint that helped dissipate heat and could stretch without flaking were both developed to cope with the stress created by the aircraft's dramatic temperature changes during operation.

2   When an aircraft or any object travels above the speed of sound it creates a shockwave, commonly known as a 'sonic boom', that is heard on the ground as a sound similar to an explosion or a clap of thunder. The sound is created when pressure waves created at the front and back of an object travelling at supersonic speeds merge to create a single shockwave – it happens not only when an object breaks the sound barrier, but is a continuous noise that follows a speeding aircraft. A great deal of research is underway to minimise the extent of sonic boom, but it cannot be eliminated entirely.

3   Lord Kelvin is often reported to have said that 'Heavier-than-air flying machines are impossible,' but there is no reliable evidence that those are his exact words. The quote I include here was from an interview with the *Newark Advocate*, 26 April 1902, p. 4.

4   In 2015 there were an estimated 263 million cars registered in the US and 218 million registered licence holders – see https://www.statista.com/statistics/183505/number-of-vehicles-in-the-united-states-since-1990/

5   For a rounded and well-researched summary of the data about international trade, see Esteban Ortiz-Ospina and Max Roser, 'International Trade', 2018. Published online at https://ourworldindata.org/international-trade

6   This research paper describes the archaeological evidence supporting the idea that northern Australia was settled by modern humans at least 65,000 years ago. C. Clarkson, Z. Jacobs, B. Marwick, R. Fullagar, L. Wallis, M. Smith, R. G. Roberts, E. Hayes, K. Lowe, X. Carah and S. A. Florin, 'Human occupation of northern Australia by 65,000 years ago', *Nature*, 547(7663), 2017, p.306.

7   John Kelly, *The Great Mortality: An Intimate History of the Black Death, the Most Devastating Plague of All Time*, Harper, New York, 2005.

8   It is difficult to discern the precise origin of the stirrup; it is probable that simple leather loops were in use at earlier dates, but the first unmistakable representation of true riding stirrups comes from a tomb near Nanjing that has been dated to AD 322. See A. E. Dien, 'The Stirrup and Its Effect on Chinese Military History', *Ars Orientalis*, 16, 1986, pp. 33–56.

9   This theory was first put forward by American historian Lynn White Jr. Some have disputed aspects of the argument since it was proposed, but the stirrup was clearly part of the system that allowed cavalry to play a dominant part in warfare until the twentieth century. See Lynn White Jr., *Medieval Technology and Social Change*, Oxford University Press, Oxford, 1964.

10 E. R. Kilby 'The Demographics of the US Equine Population', in Deborah Salem and Andrew Rowan (Eds), *The State of the Animals*, Humane Society Press, Washington, DC, 2007, pp. 175–205

11 In principle, the lack of air resistance in a Hyperloop could facilitate travel at very high speeds, up to several times greater than the speed of sound, with relatively high power efficiency. The latest incarnations of this type of scheme, including one that Elon Musk proposed in 2012, are developments of the plans put forward by rocketry pioneer Robert Goddard in 1904. The idea of using a partial vacuum to transport people or freight has an even longer history. During the nineteenth century, various engineers built 'atmospheric railways', including Isambard Kingdom Brunel's South Devon Railway, which was constructed during the 1840s. However, a range of technical challenges, including rats gnawing through leather seals, meant that the vacuum propulsion system was quickly abandoned. Even today the engineering challenge of making a safe, reliable and economically viable Hyperloop system is formidable.

12 This intriguing story is told in more detail in R. Smith, 'The Japanese Shinkansen: Catalyst for the Renaissance of Rail', *The Journal of Transport History*, 24(2), 2003, pp. 222–37.

13 Nicholas Crafts and Anthony Venables, 'Globalization in History. A Geographical Perspective', in *Globalization in Historical Perspective*, University of Chicago Press, Chicago, 2003, pp. 323–70

14 In order for an aircraft to rise into the air, it must create a lift force that exceeds the force of gravity. In aircraft that are heavier than air, lift is

generated by the flow of air over the wings. Bernoulli's equation, made in the middle of the eighteenth century, states that when the velocity of a fluid increases, its pressure must decrease. The cross-sectional profile of aircraft wings exploits this phenomenon, because the surface area of the top of the wing is greater than the bottom. As moving air flows over the top of the wing its velocity increases, which in turn decreases air pressure locally. The air pressure beneath the wing is therefore greater than that above, which generates lift.

15   Fred Howard, *Wilbur And Orville: A Biography of the Wright Brothers*, Ballantine Books, New York, 1988.

16   As Robins explained, 'Dr Griffith was the head of the Royal Aircraft Establishment, and he was working on jet engines, and he rightly saw that ultimately an axial flow engine would be the most efficient. Whittle, being a more practical guy, saw that to make a jet engine fairly quickly, a centrifugal design – which would really be like a big supercharger – would be lower risk.' Sure enough, all the early jet engines adopted the centrifugal scheme. Today however, the majority of jet engines have an axial compressor.

17   The key to making jet engines more efficient has been a steady increase in their 'bypass ratio'. In these engines, a proportion of the air entering the engine is diverted and accelerated around the outside of the main core of the engine. Thrust comes from a combination of this outer flow and the hot exhaust from the central turbine. Bypassing a greater portion of air leads to better efficiency but it also creates more drag, more weight and subjects the blades of the engine's fans to increased mechanical stress. It is what Robins calls, in his understated way, 'a very interesting trade-off'. Over the decades, the bypass ratio has grown from 0.3:1 in the Rolls-Royce Conway that Robins worked on in the late 1950s, to around 10:1 in today's most advanced turbofan engines. Constant innovation in mechanisms, materials and the understanding of aerodynamics have allowed these dramatic changes to take place.

18   Ian Savage, 'Comparing the fatality risks in United States transportation across modes and over time', *Research in Transportation Economics*, 43(1), 2013, pp. 9–22.

19   World Health Organization, 'The Top Ten Causes of Death', 2017. Available online at http://www.who.int/mediacentre/factsheets/fs310/en

20 S. Singh, 'Critical Reasons for Crashes Investigated in the National Motor Vehicle Crash Causation Survey', Traffic Safety Facts Crash Stats, National Highway Traffic Safety Administration, Washington, DC, 2015. Available at https://crashstats.nhtsa.dot.gov/Api/Public/ViewPublication/812115

21 International Organization of Motor Vehicle Manufacturers, 'Annual Production Statistics', 2016. Available online at http://www.oica.net/category/production-statistics/2016-statistics

22 According to the International Energy Agency, road vehicles were responsible for 49.7 per cent of global oil consumption in 2015, up from 30.8 per cent in 1973. 'World Energy Outlook: Organisation for Economic Co-operation and Development'. Available online at https://www.iea.org/media/publications/weo/WEO2017ToC.pdf

23 J. Conti, P. Holtberg, J. Diefenderfer, A. LaRose, J. T. Turnure and L. Westfall, 'International Energy Outlook 2016, with Projections to 2040', US DoE Energy Information Administration, Office of Energy Analysis, Washington, DC. Available online at https://www.eia.gov/outlooks/ieo/pdf/transportation.pdf

24 Ibid. Summarised online at https://www.eia.gov/todayinenergy/detail.php?id=31332

25 Centre for Economics and Business Research, 'The Future Economic and Environmental Costs of Gridlock in 2030: An Assessment of the Direct and Indirect Economic and Environmental Costs of Idling in Road Traffic Congestion to Households in the UK, France, Germany and the USA', London, 2014.

26 B. K. Sovacool, 'Early modes of transport in the United States: Lessons for modern energy policymakers', Policy and Society, 27(4), 2017, pp. 411–27.

27 Camille Jenatzy's La Jamais Contente was the first car to exceed one hundred kilometres (sixty miles) per hour, in 1899.

28 'Electric Vehicles Attract Attention', New York Times, 20 January 1911.

29 International Energy Agency, 'CO2 Emissions from Fuel Combustion 2017 Overview'. Available online at https://www.iea.org/publications/freepublications/publication/CO2EmissionsFromFuelCombustion2017Overview.pdf

30 From an adjusted, real-world efficiency of just 13.1 miles per gallon in 1975 to 25.2 miles per gallon in 2017. Figures from the US Environmental

Protection Agency Report, 'Light-Duty Automotive Technology, Carbon Dioxide Emissions, and Fuel Economy Trends: 1975 Through 2017', 2018.

31 C. von Kaenel, 'Americans are Driving More Than Ever', *Scientific American*, 22 February 2016.

32 Figures from the Global Burden of Disease project. Of the approximately eighteen deaths per 100,000 people that are caused by road transport in Western Europe, 44 per cent of them are attributable to air pollution and the others to accidents. See Global Road Safety Facility, The World Bank, Institute for Health Metrics and Evaluation, 'Transport for Health: The Global Burden of Disease from Motorized Road Transport', 2014. Available online at http://www.healthdata.org/policy-report/ transport-health-global-burden-disease-motorized-road-transport

33 International Energy Agency, 'Global Electric Vehicle (EV) Outlook', 2017. Available online at https://www.iea.org/publications/ freepublications/publication/GlobalEVOutlook2017.pdf

34 The drivetrain of a vehicle is the collection of components that deliver power to the driving wheels. This does not include the engine that generates the power.

35 The clearest summary of the relative 'tank to wheels' efficiency of gasoline, hybrid and electric automobiles is provided by the US Department of Energy Office of Energy Efficiency and Renewable Energy. Available online at https://www.fueleconomy.gov/feg/atv-ev. shtml

36 Caroline Hargroves, the Technical Director of McLaren Applied Technologies, describes how motor racing has always driven forward the cutting edge of automobile technology: 'There's a lot of technology, and a lot of approaches to the data analytics and the rapid development of parts that are very relevant outside Formula One.' Many innovations, which include the disc brake, electronic traction control and the use of light, strong carbon fibre and aluminium components, started in motorsport and trickled down to consumer vehicles. These are some of the advances that have helped to make them so impressively safe and reliable. Today, Formula One is stretching the capabilities of electric propulsion; the racing cars are now driven by a hybrid combination of an internal combustion engine and an electric motor. Much of the heat from the exhaust and kinetic energy produced by braking, which would

otherwise be lost, is either recovered and stored in a battery or used immediately to increase the car's power. Meanwhile, the increasingly popular all-electric Formula E Championship is demanding improved performance from motors and batteries.

37 GPS satellites are installed in the outer Van Allen radiation belt because it gives them a geosynchronous orbit of around twelve hours – this means they follow a predictable path, with the same timing each day. The intense radiation in the Van Allen belt, caused by the high concentration of energetic charged particles that are held captive by the Earth's magnetic field, makes the engineering of robust satellites particularly challenging. Brad Parkinson described how building the first 'radiation-hardened atomic clock', crucial for accurate timekeeping, was 'darn near a bridge too far'. The Navy Research Lab (NRL) was to be the primary source but was too late and experienced on-orbit failures. Fortunately, the Air Force had a backup from Rockwell International that saved the programme.

38 I. Evtimov, K. Eykholt, E. Fernandes, T. Kohno, B. Li, A. Prakash, A. Rahmati and D. Song, 'Robust Physical-World Attacks on Deep Learning Models', *arXiv preprint arXiv:1707.08945*, 2017.

39 Automatic emergency braking (AEB) systems are already standard features on many new automobiles today, and are usually required for a 'five star' safety rating.

40 You can read more about this new approach in more detail in: S. Shalev-Shwartz, S. Shammah and A. Shashua, 'On a Formal Model of Safe and Scalable Self-driving Cars', *arXiv preprint arXiv:1708.06374*, 2017.

41 Q. Bu, 'The Most Common Job in Every State', National Public Radio, 5 February 2015.

42 P. Barter, ' "Cars are parked 95% of the time". Let's check!', *Reinventing Parking*, 22 February 2013.

43 M. Mazur, 'Six Ways Drones are Revolutionizing Agriculture', *MIT Technology Review*, 20 July 2016.

44 D. J. Bradley, 'The Scope of Travel Medicine', in *Travel Medicine: Proceedings of the First Conference on International Travel Medicine* by R. Steffen (ed), Springer Verlag, Berlin, 1989, pp. 1–9.

45 According to the 'no-teleportation principle', which stems from Heisenberg's uncertainty principle, once matter has been disassembled in order to teleport it, you could not know how to put it back together

correctly unless the original was completely destroyed – not an appealing prospect for anyone who is keen to step into a *Star Trek*-style 'transporter'.

8 DEFEND

1   Von Neumann based his advice on logical mathematical predictions from game theory. He argued that if two powerful states both had nuclear weapons, war was inevitable. In his view, the only logical action was to strike first and with devastating power. He is reported to have said, 'If you say, "Why not bomb [the Soviets] tomorrow?", I say, "Why not today?"' (quoted in C. Blair Jr., 'The Passing of a Great Mind', *Life Magazine*, 25 February 1957). Fortunately, good sense and self-preservation prevailed.

2   H. M. Kristensen and R. S. Norris, 'Worldwide deployments of nuclear weapons', *Bulletin of the Atomic Scientists*, 73(5), 2017, pp. 289–97, and S. N. Kile and H. M. Kristensen, *Trends in World Nuclear Forces*, Stockholm International Peace Research Institute, 2017.

3   This statistic and the death rates that follow it are from Steven Pinker, *The Better Angels of Our Nature*, Penguin, London, 2012; Steven Pinker, *Enlightenment Now*, Penguin, London, 2018; and Max Roser, 'War and Peace', 2016, which is available online at https://ourworldindata. org/war-and-peace. Both of these provide references to the primary sources.

4   Ibid. The statistics here are based on the standard definition of 'war' used by political scientists: a state-based conflict that results in at least 1,000 casualties in any given year.

5   Ibid.

6   An estimated 2.8 million adults die globally each year as a result of being overweight or obese. The European Association for the Study of Obesity, 'Obesity Facts and Figures', 2018. Available online at http:// easo.org/education-portal/obesity-facts-figures/

7   Joseph Needham et al., *Science and Civilisation in China: Volume V, Part 7*, Cambridge University Press, Cambridge, 1987. The Chinese word for gunpowder, 'huǒyào', means 'fire medicine', revealing the original motivation of the alchemists who discovered it.

8   Ibid.

9   Mariana Mazzucato, *The Entrepreneurial State: Debunking Public vs. Private Sector Myths*, Anthem Press, London, 2015.

10 In 1482 the thirty-year-old Leonardo da Vinci left his native Tuscany to move to Milan. Looking for gainful employment as a military engineer, he wrote to the Duke of Milan Ludovico Sforza, boasting about his engineering prowess. 'When a place is besieged I know how to cut off water from the trenches and construct an infinite variety of bridges, mantlets and scaling ladders and other instruments pertaining to sieges. I also have types of mortars that are very convenient and easy to transport ... when a place cannot be reduced by the method of bombardment either because of its height or its location, I have methods for destroying any fortress or other stronghold, even if it be founded upon rock ... If the engagement be at sea, I have many engines of a kind most efficient for offence and defence, and ships that can resist cannons and powder.' At that stage, his ideas were still notional. He had never experienced war or actually built any of his military inventions.

11 His designs were never implemented, although modern reconstruction of his diving apparatus showed that his pigskin suit, mask and bamboo cane breathing tubes that reached to the surface could have worked. See Kenneth D. Keele, *Leonardo da Vinci's Elements of the Science of Man*. Academic Press, Cambridge, 2014.

12 Dead reckoning is a navigational technique that allows a navigator to use their speed of travel and the location of previously known geographical position to calculate current location. It is accurate over short distances, but errors creep in as distance from the known position increases.

13 Since the Earth revolves through 360 degrees every twenty-four hours, each hour is equivalent to the progression of fifteen degrees of longitude. Longitude can, therefore, be calculated easily if a ship knows the time difference between its current position and port of origin.

14 Newton is quoted in Dava Sobel, *Longitude: The True Story of a Lone Genius Who Solved the Greatest Scientific Problem of his Time*, Bloomsbury Publishing USA, 2007.

15 R. T. Gould, 'John Harrison and his timekeepers', *The Mariner's Mirror*, 21(2), 1935, pp. 115–39.

16 Winston Churchill, *The Second World War, Volume 2: Their Finest Hour*, Houghton Mifflin, Boston, 1949.

17 Werner von Braun's story is described in detail in Deborah Cadbury, *Space Race: The Epic Battle Between America and the Soviet Union for Dominion of Space*, Harper Collins, London, 2007.

18 W. von Braun, 'Reminiscences of German rocketry', *Journal of the British Interplanetary Society*, 15(3), 1956.

19 Von Braun eventually became something of a hero in America, even presenting children's television programmes about space for Walt Disney. You can watch one of these online at https://www.youtube.com/watch?v=2fautyLuuvo

20 The fission of uranium atoms was first demonstrated by Otto Hahn, Lise Meitner and Fritz Strassmann, working in Berlin in 1938. Described in J. E. Fergusson, 'The history of the discovery of nuclear fission', *Foundations of Chemistry*, 13(2), 2011, p. 145.

21 Alfred Hershey and Martha Chase proved that DNA is the genetic material in 1952, by marking the DNA of viruses with radioactive phosphorus and then showing that it is the DNA of the virus, not the protein, that carries genetic information into the cell that it infects. A. Hershey and M. Chase, 'Independent functions of viral protein and nucleic acid in growth of bacteriophage', *The Journal of General Physiology*, 36(1), 1952, pp. 39–56

22 J. von Neumann, first draft of a report on the EDVAC, *IEEE Annals of the History of Computing*, 1993, 15(4), 1945, pp. 27–75.

23 Moniz describes the Nuclear Threat Initiative as 'more of a "do tank" than a think tank'. Its mission is to 'prevent catastrophic attacks with weapons of mass destruction and disruption – nuclear, biological, radiological, chemical and cyber'.

24 K. Harnett, 'Computer Scientists Close in on Perfect, Hack-Proof Code', *Wired*, 23 September 2016. Also the DARPA program: https://www.darpa.mil/program/high-assurance-cyber-military-systems

25 M. W. Lewis, 2013. 'Drones: Actually the Most Humane Form of Warfare Ever', *Atlantic*, 21 August 2013.

26 S. McCammon, 'The Warfare May Be Remote But The Trauma Is Real', National Public Radio, 24 April 2017. Also J. L. Otto and B. J. Webber, 'Mental health diagnoses and counseling among pilots of remotely

piloted aircraft in the United States Air Force', *Medical Surveillance Monthly Report*, 20(3), 2013, pp. 3–8.

27 The letter, coordinated by MIT professor Max Tegmark, has been signed by many of the leaders in the field of AI research, including Demis Hassabis and Mustafa Suleyman of Google DeepMind, Peter Norvig and Geoff Hinton of Google, Facebook's Yann LeCun and Dileep George of Vicarious, among many others. You can read it at https://futureoflife.org/open-letter-autonomous-weapons

28 The Avtomat Kalashnikova-47, better known as the AK-47, is an iconic weapon, and a good candidate for the most unpleasant and disruptive innovation in history. First produced in 1947, it was the product of a secret Soviet competition to design a new and lethal assault rifle. At first, military experts scoffed at the AK-47 – it was inaccurate, inelegant and its bullets had less stopping power than many other contemporary rifles. However, the AK-47 soon showed its real power; the simple design means that it is almost indestructible. It continues to fire when coated in grit and mud and is so easy to use that even an untrained child soldier can use it to kill. Crucially, the AK-47 is easy and cheap to manufacture. There are at least 75 million Kalashnikovs around the world today – almost 20 per cent of all guns. No firearm has killed more people.

29 'Slaughterbots' by Ban Lethal Autonomous Weapons, 2017. Available online at https://autonomousweapons.org

30 The Kalashnikov Corporation is, reportedly, forging ahead with the development of mobile, autonomous, land-based robotic weapons. See M. Smith, 'Is "killer robot" warfare closer than we think?', *BBC News*, 25 August 2017.

31 Cummings recently wrote a report for the Chatham House think tank (M. Cummings, *Artificial Intelligence and the Future of Warfare*, Chatham House for the Royal Institute of International Affairs, 2017) in which she outlines much of the evidence that suggests that the centre of gravity in research and development funding is shifting away from state-level defence expenditure towards commercial and industrial investment.

32 See H. Sender, 'US defence: Losing its edge in technology?', *Financial Times*, 4 September 2016 and Adam J. Harrison and Paul Horn,

'National Security Technology Accelerator: A Plan for Civil-Military Industry Innovation', New York University, New York, 2015.

33 D. Frum, 'Mutually Assured Disruption', *New York Times*, 10 October 2006.

34 This does, of course, depend on where you happen to live. While warfare is broadly on the decline, there are still too many parts of the world, many of them concentrated in the Middle East, where brutal military conflicts persist and whole populations live in continual danger.

9 SURVIVE

1 In a conversation we had before the operation, Vale's colleague Ara Darzi, one of the pioneers of robotic surgery, described human hands as 'the best designed surgical instruments you can find'. The late twentieth-century development of laparoscopic (keyhole) surgery was revolutionary, because many operations became much less invasive and traumatic for the patient. The chief drawback, Darzi says, was that 'we actually substituted the human hand ... with fairly primitive tools that reduced the freedom of movement'. The da Vinci robotic surgery system gives surgeons that freedom back.

2 Figures quoted are for average global life expectancy in 2018, which hides wide variations across different countries – from Japan, where life expectancy at birth is eighty-four years, to Sierra Leone, where it is fifty-two. Data from Max Roser, 'Life Expectancy', 2018. Available online at https://ourworldindata.org/life-expectancy, which summarises historical data from the World Bank and World Health Organization. Also James C. Riley, 'Estimates of Regional and Global Life Expectancy, 1800–2001', *Population and Development Review*, 31(3), pp. 537–43, September 2005. For a series of powerful visualisations of the changes in life expectancy around the world and across time, in addition to trends in many other metrics, see https://www.gapminder.org/tools

3 World Health Organization fact sheet, 'The Top Ten Causes of Death', 2018. Available online at http://www.who.int/news-room/fact-sheets/detail/the-top-10-causes-of-death

4 H. Marsh, 'It seems the present government is content to let the NHS slowly wither', *Guardian*, 7 July 2017.

5   Christopher Hamlin, *Cholera: The Biography*, Oxford University Press, Oxford, 2009.

6   The poor were the most deeply embedded in this teeming squalor. It must have seemed self-evident that they inhaled more of the miasma and therefore laboured under the greatest burden of disease; sure enough, the frequent epidemics of cholera and other infectious disease – tuberculosis, typhoid, typhus, diphtheria, pneumonia and influenza – kept many of the labouring poor away from work.

7   However, there was no abundance of compassion here – the chief arguments were economic. The poor needn't be emancipated, but they had to be productive. The state wanted to make sure that workers were well enough to get to work, to keep driving the economy and sustain Britain's empire-building aspirations. The person who worked hardest to emphasise the causal links between filth, disease, unemployment and death was Edwin Chadwick – the Public Health Act in 1848 provided the platform from which Chadwick could start trying to rid London of her dangerous miasma.

8   Although the handle on the Broad Street water pump in Soho was removed after Snow presented his evidence, it was replaced soon thereafter – without any attempt having been made to improve the quality of the water that the pump produced.

9   Huge pumping stations, housing powerful steam engines, pushed the refuse on its way. These buildings display the optimism and confidence of the engineer. Architectural historian Nikolaus Pevsner described them as 'exciting architecture applied to the most foul purposes' – he pointed out 'an unorthodox mix, vaguely Italian Gothic in style but with tiers of Byzantine windows and a central octagonal lantern that adds a gracious Russian flavour'. Because the real engineering triumph was almost entirely buried and invisible beneath London's streets, these pumping stations were highly decorated, to create a focal point to celebrate Bazalgette's achievement. Although his project was not completely finished until 1875, there was an official opening ceremony in April 1865. The royal party were most impressed by a vast holding tank at the east end of the sewer network, 'An underground Valhalla, where the streets of columns, lights and arches spread around in all directions.' Some may have been less pleased to hear that a single sluice

gate was all that kept 'five acres of the filthiest mess in Europe' from where they stood. *The Times*, 5 April 1865

10   The last major cholera outbreak struck London in 1865–6. It was mainly confined to a section of east London that hadn't yet been plumbed into Bazalgette's sewerage system – the cause of the epidemic was traced to one particular water supply, which convinced many that John Snow had been quite right about the transmission of cholera more than a decade earlier. Snow had died in 1858 and so didn't survive to see his ideas widely accepted.

11   'Death of Sir Joseph Bazalgette', *The Times*, 16 March 1891, p. 4.

12   Today it is clear that London has pushed Bazalgette's system too far – the city is once again polluting its river with raw sewage. Bazalgette factored in the population doubling, to 4 million, but could not have predicted that the city would swell to its current 8.6 million. Nowadays, any moderate rainfall causes the sewers to overflow, spewing their contents directly into the Thames. Thirty-nine million tonnes of raw sewage enter the Thames in an average year. Once again engineers are sorting out the mess – the twenty-first-century solution is to build a vast tunnel, deep under the Thames itself. Tunnel boring machines will chew their way through London's substratum, creating a 7.2-metre diameter tunnel that will go from thirty metres deep at Acton to sixty-five metres deep in Abbey Mills in the east. Phil Stride, of Tideway, the engineering company leading the work, tells me that 'we will go under virtually all of London's infrastructure, including forty-two existing tunnels'. The Thames Tideway Tunnel should be open by 2024. Costing £4.2 billion, it is the largest privately financed infrastructure project in Europe. Stride points out that this new 'super-sewer' doesn't upstage Bazalgette: 'This is only an extension [to Bazalgette's sewer system], in terms of picking up what overflows from it.'

13   The World Bank defines 'improved water sources' as 'piped water on premises: piped household water connection located inside the user's dwelling, plot or yard', and 'other improved' drinking water sources as 'public taps or standpipes, tube wells or boreholes, protected dug wells, protected springs, and rainwater collection'. This is not synonymous with 'safe' or 'potable', although improved water sources are much more likely to be adequately separated from sewage and other sources of contamination. Hannah Ritchie and Max Roser, 'Water

Access, Resources and Sanitation', 2018. Available online at https://ourworldindata.org/water-access-resources-sanitation' and 'Progress on sanitation and drinking water: 2015 Update and MDG assessment', World Health Organization, Geneva, Switzerland.

14 Santiago Septien, 'How do we solve Africa's Sanitation Problems?' World Economic Forum, 2015. Available online at https://www.weforum.org/agenda/2015/08/how-do-we-solve-africas-sanitation-problems/

15 See Note 16.

16 Richard Hollingham, *Blood and Guts: A History of Surgery*, BBC Books, London, 2008, p. 62.

17 The importance of hand-washing was first formally demonstrated by Hungarian-born physician Ignaz Semmelweis, when he was working in a hospital in Vienna at the end of the 1840s. He noticed that the incidence of deaths caused by puerperal fever was higher among mothers of babies in wards staffed by physicians than in a nearby ward run by midwives. Semmelweis realised that the crucial difference was that the physicians conducted autopsies on patients who had been killed by infections, and so sometimes passed disease-causing agents – this was before acceptance of germ theory, so he described them as 'cadaverous particles' – to healthy patients. Rigorous hand-washing, in a solution of chlorine, dramatically reduced the rate of infection on the maternity wards, though the Viennese medical establishment were regrettably resistant to Semmelweis's suggestion that they should improve their hygiene.

18 James C. Riley, *Rising Life Expectancy: A Global History*, Cambridge University Press, New York, 2001.

19 T. W. Geisbert, 'First Ebola Virus Vaccine to Protect Human Beings?' *The Lancet*, 389(10068), 2017, pp. 479–80.

20 Figures from the World Health Organization, accessed in January 2018, who report over 28,000 cases of Ebola and 11,300 fatalities in Guinea, Liberia and Sierra Leone. Data from http://apps.who.int/gho/data/view.ebola-sitrep.ebola-summary-20160511?lang=en

21 Jeremy Farrar, Director of the Wellcome Trust, emphasises that other technologies and systems were also developed and honed to tackle the 2014–16 West African Ebola outbreak. For example, handheld DNA sequencing devices, which I will describe later in this chapter, allowed the disease to be diagnosed, patients triaged and the epidemic to be

tracked, nearly in real time. These techniques will be invaluable in the event of a new epidemic.

22 Fleming made his famously serendipitous discovery in 1928, yet no one was successfully treated with penicillin until 1942. After he had identified a substance secreted by a green-blue mould that could kill bacteria, Fleming wrote up his findings and moved on, apparently missing the huge potential that the drug had as a human therapy. He was probably not even the first to discover penicillin; research as far back as the 1890s described the anti-bacterial properties of moulds of the genus *Penicillium*. It was only as the Second World War loomed on the horizon that Fleming's research was dug out and pushed forward. Howard Florey, Ernest Chain and Norman Heatley drove the second wave of penicillin research. These biochemists isolated the active ingredient and verified its antibiotic properties. Even so, making enough to treat human patients proved beyond the academics. By 1941, the war effort squeezed the Oxford team's resources, forcing them to look to America for help. They took their penicillin-producing cultures to the US Department of Agriculture's research lab in Peoria, Illinois. It was there that the scientists passed the baton on to the engineers – it became their job to mass-produce penicillin for the war effort.

23 American Chemical Society, 'The Discovery and Development of Penicillin 1928–1945', commemorative booklet produced by the National Historic Chemical Landmarks programme of the American Chemical Society, 1999.

24 Claire Panosian Dunavan, 'The Drug that Helped Win the War', *Los Angeles Times*, 11 April 2004.

25 Until then, others had kept the fungus in its favoured habitat: growing as a thin, minimally productive film at the surface of a liquid.

26 Guru Madhavan, *Think Like an Engineer*, Oneworld Books, London, 2016.

27 These are microbes that most of us coexist with every day – they live in our intestines and mouths and on our skin. Sometimes, when our defences are down, *Klebsiella* can cause pneumonia or urinary tract infections, but ordinarily they don't cause us trouble. If they do, a course of antibiotics is usually effective.

28 Alexander Fleming, 'Penicillin', Nobel Lecture, 11 December 1945. Despite Fleming's exhortation, the misuse of antibiotics has been

widespread for decades. Too many people take them when they do not need them, do not take enough or stop their courses too early. It is also common practice in industrial agriculture to feed livestock a continuous low dose of antibiotics – the animals thrive, but so do any bacteria that acquire the genetic mutations giving them resistance to the antibiotics. Having acquired resistance, bacteria swap genes among themselves, spreading resistance further and faster.

29 The UK government commissioned a major report on antimicrobial resistance, led by Jim O'Neill, which reported its findings in 2016. J. O'Neill, 'Review on Antimicrobial Resistance', *Antimicrobial resistance: tackling a crisis for the health and wealth of nations,* 2014. Available online at https://amr-review.org

30 The crisp images produced by an MRI scanner rely on a deep understanding of quantum physics. Since the fundamental atomic particles of matter have an inherent polarity, they respond to strong magnetic fields. When a patient is placed inside the scanner, the nuclei of hydrogen atoms within her body are impelled to respond to a strong magnetic field, which causes them to line up neatly along the magnetic axis. The scanner then releases a pulse of radio energy at a precise frequency. A small subset of hydrogen atoms responds, flipping into a higher energy state known as 'resonance'. When the radio pulse is turned off, the resonant atoms return to their starting state, and as they do so, they release a small amount of energy. The MRI scanner picks up these miniscule signals. By reconstructing the timing and distribution of these tiny signals, it can build up a view of the structural details and water content of the body.

31 Another challenge here is the fact that most materials have an extremely low heat capacity at four kelvin – put simply, energy passes straight through them. To counter this, most MRI scanners contain a thermal shield between the patient and the magnet, which is itself cooled to fifty kelvin and coated in space blanket materials. The MRI scanner's superconducting coils are also suspended inside the scanner on low-conductivity carbon fibre straps, insulating them further from heat conductance. Furthermore, all the different layers of the magnet's setup are held in a complete vacuum, to minimise heat convection. Finally, a power-hungry refrigeration unit prevents re-condensation of the liquid helium that maintains the superconductive temperature.

32 The company that Calvert works for, Siemens Magnet Technology, released the first commercially available 7-tesla MRI system. Calvert called it the 'pinnacle of our achievement'. As well as matching the strongest magnets in the industry, they achieved that with much less material: their scanner is twenty-three tonnes lighter than their competition, which means Siemens manufacture the only super-high-resolution MRI machines that can be airfreighted to where they are needed in a super-cooled and charge-loaded state. However, they are still very bulky and expensive machines, which prevents widespread adoption. Meanwhile, even more powerful machines are under development, including a 21.1-tesla system that has been used to image live animals.

33 K. A. Wetterstrand, 'DNA Sequencing Costs: Data from the NHGRI Genome Sequencing Program (GSP)', the National Human Genome Research Institute, 2018. Available online at www.genome.gov/sequencingcostsdata.

34 Nanopore's ability to sequence in near real time is unusual. The more widely used second-generation 'sequencing by synthesis' approach used by market-leaders Illumina has the capacity to sequence many samples in parallel, but it takes three minutes to read each successive DNA letter.

35 Vast swathes of your genome is made up of so-called 'junk DNA', sequences that are mainly the relics of ancient viral attacks and have no clear function. As a result, variants in these segments usually have no obvious effect at all.

36 Other promising large-cohort studies around the world include All of Us in the US, the Kadoorie Biobank in China, the Mexico City Prospective Study in Mexico and the Chennai Prospective Study in India.

37 The intelligent knife (or 'iKnife') is an invention that extends the capabilities of a standard electrosurgical tool (widely used devices that use electricity to heat tissues to make clean incisions), by providing the ability to analyse the chemical composition of smoke from the cut tissue. This is done by a technique called Rapid Evaporative Ionisation Mass Spectrometry (REIMS), which can detect chemical differences between cancerous and non-cancerous cells in real time. Read more in J. Balog, L. Sasi-Szabó, J. Kinross, M. R. Lewis, L. J. Muirhead, K. Veselkov, R. Mirnezami, B. Dezső, L. Damjanovich, A. Darzi

and J. K. Nicholson, 'Intraoperative tissue identification using rapid evaporative ionization mass spectrometry', *Science Translational Medicine*, 5(194), 2013.

38 S. Prokesch, 'The Edison of Medicine', *Harvard Business Review*, March–April 2017.

39 These particles can resist the heat of boiling water, but as soon as they hit the acidic environment of the stomach they release the essential nutrients that they carry.

40 Siddhartha Mukherjee, *The Emperor of All Maladies: A Biography of Cancer*, Fourth Estate, London, 2011.

41 On a given cancer, at a given stage, many chemicals will be completely ineffective, and some will make a bad situation even worse. Even faced with two patients presenting with what looks like the same type of tumour, the disease can have completely distinct underlying genetic causes, and therefore may need very different treatments.

42 William C. Moloney and Sharon Johnson, *Pioneering Hematology: The Research and Treatment of Malignant Blood Disorders – Reflections on a Life's Work*, Francis A. Countway Library of Medicine, Boston, 1997.

43 Polaris Partners, a venture capital firm that have funded several of Langer's commercial ventures, calculate that his inventions have the potential to improve the lives of 4.7 billion people around the world. S. Prokesch, 'The Edison of Medicine', *Harvard Business Review*, March–April 2017.

44 In 2016, CRISPR-Cas9 genome editing was applied, for the first time, to a human patient, by a team at Sichuan University's West China Hospital. Their target was an aggressive lung cancer that had resisted chemotherapy and radiotherapy treatment. They took some blood, kept the immune cells from the sample alive in an incubator and then snipped out a gene called PD-1, before injecting the edited cells back into the patient. They hope that by removing the PD-1 gene, the re-engineered cells will ignore the tumour's attempts to block the immune system and will then recognise cancerous cells as dangerous intruders and destroy them. As I write in 2018, the PD-1 trial is still underway. This is a Phase 1 clinical trial – it involves just ten patients and is geared to assess the safety, rather than the efficacy of the gene editing procedure. See D. Cyranoski, 'CRISPR gene-editing tested in a person for the first time', *Nature*, 539(7630), 2016.

45 H. Wu and C. Cao, 'The application of CRISPR-Cas9 genome editing tool in cancer immunotherapy', *Briefings in Functional Genomics*, 2018.

46 Late in 2018, news broke that He Jiankui, a researcher at Southern University of Science and Technology in Shenzhen, China, had apparently used CRISPR-Cas9 gene editing to alter the DNA of two human babies for the first time. His claims were greeted with widespread condemnation and, as this book went to press, the veracity of his assertions had not been confirmed.

47 In its native state, CRISPR-Cas9 is part of a bacterial immune system – when a bacterium is invaded by a viral parasite, it has devised a way to chop up and store chunks of the virus's DNA in its own genome. The CRISPR-Cas9 system then uses this to recognise and neutralise a repeat attack from the same type of virus, while viruses avoid detection by changing their DNA sequence each time they replicate. CRISPR-Cas9 has therefore evolved to cut the DNA even when its target match is imperfect, giving the system a degree of in-built 'sloppiness'. By learning how CRISPR-Cas9 works at the molecular level, biomedical engineers have re-engineered the enzyme to increase its accuracy.

10 IMAGINE

1 According to the standard model of particle physics, all fundamental particles are given their mass by the Higgs energy field. The transfer of energy between the Higgs field and fundamental particles, which lends them mass, is mediated by particles called Higgs bosons, via a process known as the Higgs mechanism. Peter Higgs and Francois Englert were given the 2013 Nobel Prize in physics for 'for the theoretical discovery of a mechanism that contributes to our understanding of the origin of mass of subatomic particles, and which recently was confirmed through the discovery of the predicted fundamental particle, by the ATLAS and CMS experiments at CERN's Large Hadron Collider'.

2 The wavelength of a beam of electrons is as much as 100,000 times shorter than that of visible light. As a result, the resolution of the electron microscope far exceeds that of traditional light microscopes. Even the very best light microscopes are blind to structures smaller than 300 nanometres across, which is a particular problem when studying the brain. The connections between neurons, called synapses, are critical points in the brain's information-processing networks;

counting and measuring synapses is key to an understanding of how the brain works. Most synapses, however, are less than fifty nanometres across, out of reach of light microscopy.

3   The situation is further complicated by the additional presence of an even greater number of glial cells and microscopic blood vessels that, together with the tangles of neural connections, make the brain one of the most complex structures imaginable. For estimates of numbers of neurons and other cells in the brain and their connections, see F. A. Azevedo, L. R. Carvalho, L. T. Grinberg, J. M. Farfel, R. E. Ferretti, R.E. Leite, R. Lent and S. Herculano-Houzel, 'Equal numbers of neuronal and nonneuronal cells make the human brain an isometrically scaled-up primate brain', *Journal of Comparative Neurology*, 513(5), 2009, pp. 532–41.

4   Boyden qualified this statement by saying: 'I mean the published worm network is, I believe, a pastiche of six different worms. So it may not actually correspond to a working nervous system.'

5   Method first published in 2015. See F. Chen, P. W. Tillberg and E. S. Boyden, 'Expansion microscopy', *Science*, 347(6221), 2015, pp. 543–48. Improved and extended in J. B. Chang, F. Chen, Y. G. Yoon, E. E. Jung, H. Babcock, J. S. Kang, S. Asano, H. J. Suk, N. Pak, P. W. Tillberg and A. T. Wassie, 'Iterative expansion microscopy', *Nature Methods*, 14(6), 2017, p. 593.

6   A major study conducted by the World Health Organization showed that at least 300 million people, the equivalent of 4.4 per cent of the world's population, suffer from depression worldwide. It is emphatically not a first-world problem. Poverty, unemployment and major negative life events, such as the death of loved ones, exacerbate depression. Anxiety disorders, meanwhile, are estimate to affect over 264 million people globally. Across the world, depression is the leading cause of disability, as measured by 'years lived with disability', and anxiety disorders the sixth. 'Depression and Other Common Mental Disorders: Global Health Estimates', World Health Organization, 2017. Available online at http://apps.who.int/iris/bitstream/10665/254610/1/WHO-MSD-MER-2017.2-eng.pdf?ua=1

7   More information about these programmes is available online at https://www.darpa.mil/program/systems-based-neurotechnology-for-emerging-therapies

8   C. Funk, B. Kennedy and E. Sciupac, 'US Public Wary of Biomedical Technologies to "Enhance" Human Abilities', *Pew Research Center*, 2016, pp. 1–131.

9   In crude terms, information transmission is very much faster in computer microchips than in the brain. Signals travel along neurons at 120 metres per second, whereas they move at near light speed in an integrated circuit. The 'clock rate' of the human brain (one measure of processing speed) has been estimated to be around one kilohertz (similar to the first digital computers), while the latest microprocessors run at several gigahertz, more than a million times faster. And whereas our working memory can only handle four or five 'chunks' of information at any one time, today's computers can draw on multiple gigabytes of random access memory (RAM). Despite losing out in these bald comparisons, our brains display a much higher degree of parallel processing and a more sophisticated architecture than any silicon computer. They are also very much more energy-efficient, running on approximately twenty watts (less than most lightbulbs). It has been estimated that an equally capable computer would require an impractical one gigawatt of power, the equivalent of 500 utility scale wind turbines.

10 Hugh Sebag-Montefiore, *Enigma: The Battle for the Code*, Weidenfeld & Nicolson, London, 2011.

11 Good presented his ideas in public several times before writing them up. I. J. Good, 'Speculations concerning the first ultra intelligent machine', *Advances in Computers*, Volume 6, Elsevier, 1966, pp. 31–88.

12 Nick Bostrom, *Superintelligence: Paths, Dangers, Strategies*, Oxford University Press, Oxford, 2014.

13 R. Cellan-Jones, 'Stephen Hawking warns artificial intelligence could end mankind', *BBC News*, 2 December 2014.

14 L. Floridi, 'Should we be afraid of AI?', *Aeon*, 9 May 2016.

15 Michael Jordan, 'Artificial Intelligence – The Revolution Hasn't Happened Yet', *Medium*, 19 April 2018.

16 L. Lucas and R. Waters, 'The AI arms race: China and US compete to dominate big data', *Financial Times*, 1 May 2018.

17 The robot used its natural language processing skills and its ability to draw on medical knowledge to pass the written exam. See Ma Si

and Cheng Yu, 'Chinese robot becomes world's first machine to pass medical exam', *China Daily*, 10 November 2017.

18   Lagrange Points are positions in space where gravity from the sun and Earth balance the orbital motion of a satellite. The result is that a satellite placed at one of these points maintains a fixed position relative to the sun and Earth and only needs minimal thruster activity to maintain its orientation at that position. There are five Lagrange Points. The Second Lagrange Point (or L2) will be home to the James Webb Space Telescope. It is approximately 930,000 miles away from the Earth, in the opposite direction from the Sun. In this position, the telescope will have an uninterrupted view of the universe and, since it is so far into deep space, shielding it from the sun's radiation will be easier.

19   Due to the constant expansion of the universe, the very oldest stars are also likely to be furthest away from us. The increase in relative distance between us and these ancient stars causes a dramatic shift in the wavelengths of the light that is emitted from these stars as it travels to reach us. This is called a redshift – it is a similar phenomenon to the Doppler Effect, which explains why the pitch of sound changes when two objects move relative to each another. The cosmological redshift means that the oldest and most distant astronomical objects can only be detected at wavelengths that are longer than visible light. For this reason, most of the James Webb Telescope's detectors are tuned to infrared spectra.

20   D. B. Yaden, J. Iwry, K. J. Slack, J. C. Eichstaedt, Y. Zhao, G. E. Vaillant and A. B. Newberg, 'The Overview Effect: Awe and Self-Transcendent Experience in Space Flight', *Psychology of Consciousness: Theory, Research, and Practice*, 3(1), 2016, p. 1.

21   Text of J. K. Rowling's speech, *Harvard Gazette*, 2008. Available online at https://news.harvard.edu/gazette/story/2008/06/text-of-j-k-rowling-speech/

22   Credit for this way of synthesising the four great revolutions in humankind's view of themselves (namely the Copernican, Darwinian, Freudian and information revolutions) goes to University of Oxford philosopher Luciano Floridi, who wrote about it here: L. Floridi, 'Turing's three philosophical lessons and the philosophy of information', *Philosophical Transactions of the Royal Society A*, 370, 1971, pp. 3536–42.

INDEX

Page numbers in bold refer to information in image captions.

Brexit referendum 101
Breydenbach, Bernhard von **185**, 186
British Museum 85, 88, 187–8, 245
Brown, Clive 263
Brown, Neave 127
bubonic plague 258
Buckle, Henry Thomas 223
buildings, height restrictions 122–9
Bukht-Yishu 245–7
bullet trains 3, 191–2
Burj Khalifa tower, Dubai 116
Bush, Vannevar 303, 341–2n19
business, and engineering 6–7
business opportunities 32
Byron, Lord 46–7

calculators 53, 341–2n19
Calvert, Simon 260
Cambridge Analytica 101
camera phones 111
cameras 111–2
Canberra 137
cancer 69, 262, 269–70, **270**, 272, 384n41
cannons 222, 226
carbolic acid 254, **255**
carbon capture 176, 178
carbon dioxide emissions 24, 40, 161, 162, **162**, 164, 164–5, 171, 172, 175, 176, 178–9, **178**, 200
carbon fibre 126
carbon taxes 173–4
Carfrae, Tristram 129–30
Carr, Nicholas 84
Cartesian geometry 86
cathedrals, Gothic 122–4, **124**
Cauchy, Augustin-Louis 68
cement 115–6, 124, 355–6n1, 356n2, 357n13
Center for Disease Control 258
central heating 142–3
Central Japan Railway 192

Centre for Synthetic Biology 22–3
Cerf, Vint 64, 95, 97, 99, 100, 101, 347n45
CERN 279
Chadwick, Edwin 378n7
Chain, Ernst 257–8
Chandratillake, Suranga 61, 95
change, speed of 2, 33, 301
charge-coupled device (CCD) 111
Chartres Cathedral 123
Chase, Martha 375n21
chemical fertilisers 31
chemotherapy 269
Chernobyl disaster 167
Chicago 126, 253–4
child mortality rates 2
childbirth 249, 254
Chile 138–9, 140, **140**, 143
chimeric antigen receptor T-cell (CAR-T) therapy 272, 384n44
China 112, 150, 187–8, 216, 266, 291
    connectivity infrastructure 109
    Cultural Revolution 220
    factories 134
    high-tech hubs 38
    invention of gunpowder 222
    invention of printing 350n8
    social credit system 103–4
    Three Gorges Dam 116
    urban planning 134–5
    wind-generating capacity 170
Chioggia, Battle of 226
cholera 252, **252**, 254
Chou, Charina 110
Church, George 273–5, 300, 334n17
Churchill, Winston 130, 155, 231
cities
    benefits 122, 128
    challenges 122
    development 121–9
    emergence of 148
    Foster's vision for 133–4

World Health Organization 198
World Wide Web 93–4, 95–6
Wren, Christopher 119–21, 126
Wright brothers 2, 194, 213–4
writing 75, 83, 84, 86
Wyss Institute for Biologically Inspired
    Engineering 273–4

X-ray lasers 279

Yakir, Dan 163–4
Yoshida, Kenichi 106

Zagros Mountains, Iran 146–7, 153
zero 86
zero-carbon economy 165, 178
Zetsche, Dieter 20–2, 201, 202–3
Zhang, Tong 73, 289, 291
Zuckerberg, Mark 101

A NOTE ON THE TYPE

The text of this book is set in Minion, a digital typeface designed by Robert Slimbach in 1990 for Adobe Systems. The name comes from the traditional naming system for type sizes, in which minion is between nonpareil and brevier. It is inspired by late Renaissance-era type.

John Browne trained as an engineer, was CEO of BP from 1995 to 2007 and remains an influential leader in the energy business. He is Chairman of the Crick Institute, a Fellow of the Royal Society, past President of the Royal Academy of Engineering and former Chairman of Tate. He is a collector of antique books and art and the author of four books.

Today's unprecedented pace of change leaves many people wondering what new technologies are doing to our lives. Has social media robbed us of our privacy and fed us with false information? Are the decisions about our health, security and finances made by computer programmes inexplicable and biased? Are robots going to take our jobs? Will better healthcare lead to an ageing population that cannot be cared for? Will we all be terrorised by autonomous drones that can identify and kill us, one by one? And has our demand for energy driven the Earth's climate to the edge of catastrophe?

John Browne argues that we need not and must not put the brakes on technological advance. Civilisation is founded on engineering innovation; all progress stems from the human urge to make things and to shape the world around us, resulting in greater freedom, health and wealth for all. Drawing on history, his own experiences and conversations with many of today's great innovators, he uncovers the basis for innovation and its consequences, both good and bad. He argues compellingly that the same spark that triggers each innovation can be used to counter its negative consequences. *Make, Think, Imagine* provides an eloquent blueprint for how we can keep moving towards a brighter future.